Ben Stacy Jerrik (Ed.)

Peter Mollez

Ben Stacy Jerrik (Ed.)

Peter Mollez

Arsenal F.C., Club Brugge K.V., K.V. Kortrijk, Football in Belgium

Part Press

Imprint

Publisher:
Part Press is a trademark of
International Book Market Service Ltd., 17 Rue Meldrum, Beau Bassin, 1713-01 Mauritius
Email: info@bookmarketservice.com
Website: www.bookmarketservice.com

Published in 2012

Printed in: U.S.A., U.K., Germany. This book was not produced in Mauritius.

ISBN: 978-613-9-37188-4

Contents

Articles

References

Peter_Mollez

Personal information			
Full name	Peter Mollez		
Date of birth	23 September 1983		
Place of birth	Poperinge, Belgium		
Playing position	goalkeeper		
Number	24		
Youth career			
1989-1993	VK Ieper		
1993-1997	KV Kortrijk		
1997-2003	Club Brugge		
Senior career*			
Years	**Team**	**Apps†**	**(Gls)†**
2003-2005	Club Brugge	0	(0)
2004-2005	→ Cercle Brugge (loan)	25	(0)
2005	AA Gent	1	(0)
2006-2009	KV Kortrijk	93	(0)
2009-2010	FCV Dender[1]	29	(0)
National team			
	Belgium	0	(0)

* Senior club appearances and goals counted for the domestic league only.
† Appearances (Goals).

Peter Mollez (born 23 September 1983 in Poperinge) is a Belgian football goalkeeper.

Career

Mollez was a promising youngster. He was even offered a test period at Arsenal, but Club Brugge refused that. However, Mollez never achieved his break-through at Club, and he was eventually loaned to Cercle Brugge. Quite ironically, Mollez made his début at the highest level of Belgian football with Cercle Brugge in a match against Club Brugge. Mollez had to replace the red carded Ricky Begeyn. Cercle lost the match 5-0.[2]

Mollez finally became a first team regular with KV Kortrijk in second division. Next season, Mollez will be seen again at the highest level, as KV Kortrijk became Second Division champions.

References

[1] Peter Mollez | FOOTGOAL (http://www.footgoal.net/fichejoueur.php?id_comp=12:&id_club=66&id_joueur=1229)
[2] "Stats Centre: Peter Mollez Facts" (http://guardian.touch-line.com/StatsCentre.asp?CTID=6&CPID=126&TEID=7451&PLID=179734&pStr=Player). Guardian.co.uk.. Retrieved 2009-11-12.

R._Charleroi_S.C.

Full name	Royal Charleroi Sporting Club
Nickname(s)	*Les Zèbres* (The Zebras), *Les Carolos*
Founded	1904
Ground	Stade du Pays de Charleroi, Charleroi (Capacity: 25,000[1])
Chairman	Abbas Bayat
Manager	Tibor Balogh
League	Belgian Second Division
2010-2011	Belgian Pro League, 16th *(relegated)*

Home colours **Away colours**

Sporting du Pays de Charleroi (often simply known as **Charleroi** or **Sporting Charleroi**, or by their nickname **Les Zèbres**) is a Belgian football club based in the city of Charleroi, in the province of Hainaut. Charleroi plays in the Belgian Pro League and their current spell at the highest level in Belgian football has started in the 1985–86 season. Charleroi was founded in 1904 and they first reached the first division in 1947–48. Their highest finish was runner-up in the 1968–69 season. They also reached twice the Belgian Cup final, losing in 1977-78 to Beveren and in 1992-93 to Standard Liège.

Sporting Charleroi have a long-standing rivalry with city other club ROC de Charleroi-Marchienne, currently playing in the third division. Charleroi play their home matches at the Stade du Pays de Charleroi, which was refurbished for the UEFA Euro 2000. The stadium hosted 3 group stage games in the Euro 2000 among which the 1-0 victory of England against Germany. Charleroi have been recruiting several French players in recent years, including Michaël Ciani, Cyril Théréau and goalkeeper Bertrand Laquait.

History

Charleroi Sporting Club was founded in 1904 and they received the matricule n°22 . Twenty years after their foundation, they qualified to play in the Promotion (then the second level in Belgian football) and in 1929, the club changed its name to Royal Charleroi Sporting Club. Rivals from Olympic Charleroi were playing in the first division in the late 1930s and the 1940s, while Sporting Charleroi was playing one level down, until they promoted in 1947. In 1949, Sporting Charleroi finished 4th (2 points behind Standard Liège) whereas Olympic Charleroi was 14th. But Olympic took the lead again until 1955 and their relegation to the second division. At the end of the 1956–57 season, Olympic Charleroi had promoted to the first division but Sporting Charleroi finished last in the first division and was thus relegated to the second division. A spell of 9 seasons in the second division followed and in 1966–67 Sporting Charleroi was back at the top level. They finished at the second place in 1968-69 5 points behind Standard Liège but within two years they were relegated again.

In 1974 the first division was changing from 16 to 20 teams and Sporting Charleroi was chosen to play at the top level. Olympic Charleroi promoted too as they had won the second division right before but they remained at the top level for just one season. Sporting underwent a new relegation in 1979–80 (17th on 18) but was back five years later. Their best result since then in the first division is a 4th place in 1993–94. In September 2005, the G-14 took FIFA to court over the 8 month injury incurred by Abdelmajid Oulmers whilst on international duty with Morocco.

Colours and badge

The colours of Charleroi are black and white with a shirt generally striped, which lead to the team being nicknamed The Zebras. In the 2005-06 season the team's away kit was pink.

Stadium

Main article: Stade du Pays de Charleroi

The actual ground was baptized in 1939 with a match Sporting-Union du Centre and it was located near the coal mine named Mambourg. In 1985 the stadium was slightly modernized as the club had qualified for the first division. It was then heavily renewed in the late 1990s in view of the 2000 European Football Championship. The name changed on May 24, 1999 from Stade du Mambourg to Stade du Pays de Charleroi. During the tournament, the full capacity of the stadium was up to 30,000 seats. The Stade du Pays de Charleroi hosted notably the match between Germany and England. The highest stand was eventually reduced and the capacity is now 25,000.

Honours

- **Belgian First Division:**
  - Runners-up (1): 1968-69
- **Belgian Second Division:**
  - **Winners (1):** 1946-47
  - Runners-up (1): 1965-66
- **Belgian Second Division Final Round:**
  - **Winners (1):** 1985
- **Belgian Cup:**
  - Runners-up (2): 1977-78, 1992–93

European record

As of December, 2008.

Competition	A	B	C	D	E	F	G
UEFA Cup	2	6	4	0	2	10	8
Intertoto Cup	3	10	3	3	4	11	11

A = appearances, B = matches played, C = won, D = drawn, E = lost, F = goals for, G = goals against.

Season	Competition	Round		Club	Home	Away
1969-70	UEFA Cup	1R		NK Zagreb	2-1	3-1
1969-70	UEFA Cup	2R		FC Rouen	3-1	0-2
1994-95	UEFA Cup	1R		Rapid Bucureşti	2-1	0-2
1995	UEFA Intertoto Cup	Group 10		Beitar Jerusalem		1-0
1995	UEFA Intertoto Cup	Group 10		Bursaspor	0-2	
1995	UEFA Intertoto Cup	Group 10		FC Košice		2-3
1995	UEFA Intertoto Cup	Group 10		Wimbledon FC	3-0	
1996	UEFA Intertoto Cup	Group 4		Silkeborg IF	2-4	
1996	UEFA Intertoto Cup	Group 4		Conwy United FC		0-0
1996	UEFA Intertoto Cup	Group 4		Zaglebie Lubin	0-0	
1996	UEFA Intertoto Cup	Group 4		SV Ried		3-1
2005	UEFA Intertoto Cup	2R		Tampere United	0-0	0-1

Players

Current squad

As of 2 July, 2011

Note: Flags indicate national team as has been defined under FIFA eligibility rules. Players may hold more than one non-FIFA nationality.

No.	Position	Player
3	DF	Elvedin Džinič
5	DF	Mourad Satli
7	FW	Hermanni Vuorinen
8	DF	Martos
9	MF	Thibaut Delplanck
10	MF	Hervé Kage
12	GK	Parfait Mandanda
13	DF	Richard Ofori *(on loan from Mighty Jets)*
14	MF	Samuel Fabris
15	GK	Adrien Faidherbe
16	MF	Pedro
17	FW	Matija Smrekar
18	FW	Moussa Gueye
21	DF	Mijuško Bojović

No.		Position	Player
23		GK	Cyprien Baguette
24		MF	Onur Kaya
25		MF	Abraham Kudemor
26		DF	Christopher Verbist
27		MF	Ederson
28		MF	Viktor Bopp
29		GK	Stéphane Coqu
30		MF	Leandro Almeida
31		FW	Harlem Gnohéré
77		MF	Danijel Miličević
80		MF	Kenneth Houdret
93		DF	Jean-Christophe Vergerolle
99		FW	Ivan Yagan *(on loan from Lokeren)*

Out on loan

Note: Flags indicate national team as has been defined under FIFA eligibility rules. Players may hold more than one non-FIFA nationality.

No.		Position	Player
22		DF	Matan Ohayon *(at Hapoel Petah Tikva)*
26		DF	Grégory Lazitch *(at Royal Excelsior Virton)*

For recent transfers, see List of Belgian football transfers summer 2011.

Noted players

Belgium

- Philippe Albert
- Roberto Bisconti
- Toni Brogno
- Laurent Ciman
- Alexandre Czerniatynski
- Frank Defays
- Philippe Desmet
- Grégory Dufer
- Raymond Mommens
- Olivier Renard
- Enzo Scifo
- Tim Smolders
- François Sterchele
- Olivier Suray
- René Thirifays
- Daniel Van Buyten
- Leo Van Der Elst
- Eric Van Meir

Algeria

- Adlène Guedioura
- Nasreddine Kraouche

Argentina

- Cristian Leiva

Australia

- Graham Arnold
- Eddie Krncevic

Bosnia and Herzegovina

- Enver Hadziabdic

Brazil

- Dante
- Eduardo

Burkina Faso

- Badou Kere

Cameroon

- Jean-Jacques Missé-Missé

Chile

- Osvaldo Hurtado
- Francisco Ugarte

Croatia

- Dražen Brnčić
- Nikola Jerkan
- Marijan Mrmić

England

- Carl Airey (1986) 3 games 0 goals
- Laurie Cunningham (1987) 1 game 0 goals
- Michael Peter Harrison (1985–87) 49 apps 1 goal
- John McGinlay (1982–84) [2]
- Kevin Pugh (1985–1992) 174 games 8 goals

France

- Fabien Camus
- Sébastien Chabaud
- Michaël Ciani
- Bertrand Laquait
- Laurent Macquet
- Loris Reina
- Cyril Théréau

Guinea

- Ibrahima Diallo
- Ousmane Bangoura
- Kanfory Sylla

Hungary

- Imre Garaba
- Miklós Lendvai
- Zsolt Petry
- István Pisont

Iran

- Hossein Sadaghiani
- Dariush Yazdani
- Alireza Emamifar
- Mohammad Reza Mahdavi
- Mehrdad Minavand

Ireland

- Gary Waddock (1987–89) 40 games 1 goal

Luxembourg

- Nico Braun

Macedonia

- Čedomir Janevski

Morocco

- Majid Oulmers

Netherlands

- Chris Dekker

Nigeria

- Joseph Akpala
- Victor Ikpeba

Poland

- Mirosław Justek
- Kazimierz Kmiecik

Romania

- Rodion Cămătaru

Serbia

- Ranko Stojić

Sierra Leone

- Ibrahim Kargbo
- Mustapha Sama

Sweden

- Lars Eriksson
- Pär Zetterberg

Togo

- Adékambi Olufadé

Ukraine

- Serhiy Serebrennikov

Oman

- Amad Al Hosni

Managers

- Aimé Anthuenis (1987–89)
- Georges Leekens (1994–95)
- Robert Waseige (1992–94, 1997–2000 and 2003–04)
- Jacky Mathijssen (2004–07 and 2010)
- Philippe Vande Walle (2007)
- Thierry Siquet (2007–08)
- John Collins (2008–09)
- Stéphane Demol (2009)
- Tommy Craig (2009–10)
- Jacky Mathijssen (2009)
- Stéphane Demol (2009)
- Jos Daerden (2011)

References

[1] "Stade" (http://www.sporting-charleroi.be/club-stade.php). .
[2] http://www.neilbrown.newcastlefans.com/player1/johnmcginley.htm

External links

- (French) Official Website (http://www.sporting-charleroi.be/hp/fr/hp.asp)
- (English) Charleroi SC at UEFA.com (http://www.uefa.com/footballEurope/Club=52886/domestic.html)
- Charleroi SC at EUFO.DE (http://www.eufo.de/football/bel/chari_sc.htm)
- Charleroi SC at Weltfussball.de (http://www.weltfussball.de/teams/sporting-charleroi/)
- Charleroi SC at Playerhistory.com (http://www.playerhistory.com/club/1010/)
- Charleroi SC at Transfermarkt.de (http://www.transfermarkt.de/de/verein/172/rsc-charleroi/uebersicht/startseite.html)
- Charleroi SC at Football Squads.com (http://www.footballsquads.co.uk/belgium/2008-2009/eerste/charleroi.htm)
- Charleroi SC at National Football Teams.com (http://www.national-football-teams.com/v2/club.php?id=229)
- Charleroi SC at Football-Lineups.com (http://www.football-lineups.com/team/Charleroi_SC/)

K.S.C._Lokeren_Oost-Vlaanderen

Full name	Koninklijke Sporting Club Lokeren Oost-Vlaanderen
Nickname(s)	*Tricolores*
Founded	22/01/1923 (creation) July 1, 2003 (last merge)
Ground	Daknamstadion, Lokeren (Capacity: 9,271)
Chairman	Roger Lambrecht
Manager	Peter Maes
League	Belgian Pro League
2010-11	Belgian Pro League, 6th
Home colours **Away colours**	

Koninklijke Sporting Club Lokeren Oost-Vlaanderen (Dutch pronunciation: [ˈkoː.ˌnɪŋk.lə.kə ˈspɔr.ˌtɪŋ ˈklʏp ˈloː.kə.rə(n) ˈoːst ˈvlaːn.də.rə(n)]) (often simply called **Sporting Lokeren** Dutch pronunciation: [ˈspɔr.ˌtɪŋ ˈloː.kə.rə(n)] or **Lokeren** Dutch pronunciation: [ˈloː.kə.rə(n)]) is a Belgian professional football club based in the city of Lokeren, in the province of East Flanders. Lokeren plays in the Belgian Pro League. The club was founded in 1923 and they first reach the first division in 1974-75. Since then, they only had a short spell in the second division between 1993-94 and 1995-96. Lokeren has been mostly successful in the 1980s, achieving a second place in the league in 1980-81 as well as a Belgian Cup final the same year. Their best European result was a quarter final at the 1980-81 UEFA Cup.

In the year 2000, the club merged with K Sint-Niklase SKE They are registered to the Royal Belgian Football Association with the matricule number 282. Lokeren colours are white, black and yellow. They play their home games at the Daknamstadion.

In 2012, Sporting Lokeren won their first prize, after beating KV Kortrijk in the Cup Final.

History

The matricule n°282 was given in 1920 to a club named Football Club Racing Club Lokeren (nicknamed Racing FC), but the team stopped its activity the next year. In 22 January 1923 Racing Club Lokeren was founded . Between 1945 and 1951, it had a slight name change (to Racing Athletiek- en Football Club Lokeren) and the new name since 1951 was Koninklijke Racing Club Lokeren. Due to financial problems, the fusion with the other team from the town (Koninklijke Standaard F.C. Lokeren) became necessary in 1970. The new club was then named Koninklijke Sporting Club Lokeren. In 2000, the club merged with Koninklijke Sint-Niklaas S.K.E. to form Sporting Lokeren Sint-Niklaas Waasland. A last name change was made in 2003.

Honours

- **Belgian First Division:**
  - **Runners-up (1):** 1980-81
- **Belgian Second Division:**
  - **Winners (1):** 1995-96
- **Belgian Cup:**
  - **Winners (1):** 2011-12
  - **Runners-up (1):** 1980-81

European record

As of December, 2008.

Competition	A	GP	W	D	L	GF	GA
UEFA Cup	6	28	11	9	8	38	29
Intertoto Cup	4	18	7	2	9	25	28

Season	Competition	Round		Club	Home	Away
1976-77	UEFA Cup	1R		Red Boys Differdange	3-1	3-0
		2R		FC Barcelona	2-1	0-2
1980-81	UEFA Cup	1R		FC Dynamo Moscow	1-1	1-0
		2R		Dundee United	0-0	1-1
		3R		Real Sociedad	1-0	2-2
		QF		AZ 67 Alkmaar	1-0	0-2
1981-82	UEFA Cup	1R		FC Nantes	4-2	1-1
		2R		Aris Thessaloniki FC	4-0	1-1
		3R		1. FC Kaiserslautern	1-0	1-4
1982-83	UEFA Cup	1R		Stal Mielec	0-0	1-1
		2R		Benfica	1-2	0-2
1987-88	UEFA Cup	1R		Honvéd Budapest	0-0	0-1
1999	UEFA Intertoto Cup	2R		IA Akranes	3-1	3-1
		3R		FC Metz	1-2	1-0
2001	UEFA Intertoto Cup	1R		B68 Toftir	0-0	4-2
		2R		Zaglebie Lubin	2-1	2-2
		3R		Newcastle United	0-4	0-1
2002	UEFA Intertoto Cup	1R		FC WIT Georgia	3-1	2-3
		2R		Vfb Stuttgart	0-1	0-2
2003-04	UEFA Cup	QUAL		Dinamo Tirana	3-1	4-0
		1R		Manchester City	0-1	2-3
2005	UEFA Intertoto Cup	1R		JK Trans Narva	0-1	2-0
		2R		BSC Young Boys	1-4	1-2

Current squad

As of January 10, 2012.

Note: Flags indicate national team as has been defined under FIFA eligibility rules. Players may hold more than one non-FIFA nationality.

No.	Position	Player
1	GK	Boubacar Barry
2	DF	Sepp De Roover
4	DF	Jérémy Taravel
5	DF	Mijat Marić
6	MF	Tiko
7	MF	Killian Overmeire
8	DF	Koen Persoons
9	FW	Hamdi Harbaoui
10	MF	Ivan Leko
11	FW	Benjamin De Ceulaer
12	GK	Jugoslav Lazić
13	DF	Georgios Galitsios
14	FW	Benjamin Mokulu

No.	Position	Player
16	MF	Jore Trompet
17	MF	Geir Ludvig Fevang
18	FW	Baye Djiby Fall
20	FW	Donovan Deekman
22	MF	Ebrahim Seedat
23	DF	Ibrahima Gueye
24	MF	Ayanda Patosi
25	FW	Bigen Lusala *(on loan from Anderlecht U19)*
26	MF	Cedric Mpembele
28	DF	Laurens De Bock
29	FW	Nill De Pauw
30	GK	Stefan Deloose

Out on loan

Note: Flags indicate national team as has been defined under FIFA eligibility rules. Players may hold more than one non-FIFA nationality.

No.	Position	Player
21	FW	Alfreð Finnbogason *(at Helsingborgs IF)*

For recent transfers, see List of Belgian football transfers summer 2010.

Famous former players

- Guy Dardenne
- Davy De Beule
- Maurits De Schrijver
- Filip De Wilde
- Raymond Mommens
- Jurgen Sierens
- René Verheyen
- Bruno Versavel
- Suvad Katana
- João Carlos
- Aristide Bancé
- Patrice Zéré
- Tomislav Sokota
- Jan Koller
- Roman Vonášek
- Daniel Zítka
- Elos Elonga-Ekakia
- Hervé Nzelo-Lembi
- Preben Elkjær Larsen
- Dale Tempest
- Marko Myyry
- Kari Ukkonen
- Sambégou Bangoura
- Souleymane Youla
- Marel Baldvinsson
- Arnar Grétarsson
- Arnór Guðjohnsen
- Rúnar Kristinsson
- Arnar Viðarsson
- Omer Golan
- Avi Strool
- Yoav Ziv
- Edward Linskens
- René van der Gijp
- Moussa Maazou
- Stephen Keshi
- Peter Rufai
- Samson Siasia
- Dawid Janczyk
- Grzegorz Lato
- Włodzimierz Lubański
- Jim Bett
- Jim Tolmie
- Zvonko Milojević
- Karol Dobiaš
- Mladen Dabanovič
- Adékambi Olufadé
- Đorđe Vujkov

External links

- (Dutch) / (French) / (English) Official Website [1]
- (English) Lokeren at UEFA.COM [2]
- Lokeren at EUFO.DE [3]
- Lokeren at Weltfussball.de [4]
- Lokeren at Playerhistory.com [5]
- Lokeren at Transfermarkt.de [6]
- Lokeren at Football Squads.co.uk [7]
- Lokeren at National Football Teams.com [8]
- Lokeren at Football-Lineups.com [9]

References

[1] http://www.sporting.be
[2] http://www.uefa.com/footballEurope/Club=51949/domestic.html
[3] http://www.eufo.de/football/bel/ksc_eren.htm
[4] http://www.weltfussball.de/teams/sporting-lokeren/
[5] http://www.playerhistory.com/club/14268/
[6] http://www.transfermarkt.de/de/verein/498/ksc-lokeren/uebersicht/startseite.html
[7] http://www.footballsquads.co.uk/belgium/2008-2009/eerste/lokeren.htm
[8] http://www.national-football-teams.com/v2/club.php?id=230
[9] http://www.football-lineups.com/team/Lokeren/

S.V._Zulte_Waregem

Full name	Sportvereniging Zulte Waregem
Nickname(s)	Essevee
Founded	1950 (creation) July 1, 2001 (merge)
Ground	Regenboogstadion, Waregem (Capacity: 8,500)
Chairman	Willy Naessens
Manager	Francky Dury
League	Belgian Pro League
2010-11	Belgian Pro League, 11th Group B Europa League Playoff, 2nd
Home colours **Away colours**	

Sportvereniging Zulte Waregem (Dutch pronunciation: [ˈspɔrt.fər.ˌeːnə.ɻɪŋ ˈzʏl.tə ˈβaː.rə.ɻɛm]), usually called **Zulte Waregem** (Dutch pronunciation: [ˈzʏl.tə ˈβaː.rə.ɻɛm]) or by their nickname **Essevee** (Dutch pronunciation: [ˈɛ.ˈseː.ˈveː]), is a Belgian professional football club based in Waregem, West Flanders. Zulte Waregem plays in the Belgian Pro League. Their highest finish at the highest level was a 5th place in Belgian First Division 2008-09. They have won 1 Belgian Cup. They subsequently qualified for the 2006-07 UEFA Cup, their only European season ever, only to lose in the round of 32 to Newcastle United.

The club is a product of the 2001 merger between Zultse VV and KSV Waregem, a former first division regular. Zulte Waregem first reached the highest level in Belgian football by winning the 2004-05 second division. The club outfits are red and green. They play their home matches at the Regenboogstadion, the former stadium of KSV Waregem.

History

Zulte Sportief was founded in 1950 and directly became a member of the national association (matricule n° 5381). The club **Zultse V.V.** (with the same matricule) was started after the 1976 merger between Zulte Sportief and S.K. Zulte. In 2001, the team merged with K.S.V. Waregem, who had played in the first division for many years. No team from Zulte ever achieved promotion to the second division before **S.V. Zulte Waregem** in 2002. Zultse V.V., though, gained access to the third division in 1995. The club finished 14th out of 16 for two seasons and was relegated the second time after the playoff.

Zulte Waregem playing Newcastle United in a UEFA Cup tie.

In 1999, the season in which it came back, Zultse finished 4th in this division. Two years later, the new club became champion of the 3rd division A. The registered office of Zulte Waregem lies in Zulte but the club is based in the stadium of K.S.V. Waregem in Waregem and the red and white colours of K.S.V. Waregem prevail.

The club got a chance to shine on the European stage during the 2006/07 season, reaching the Round of 32 of the UEFA Cup, facing off against English club Newcastle United, although they lost 4-1 on aggregate.

Stadium

Zulte Waregem plays its home matches at the Regenboogstadion, which means Rainbow Stadium in Dutch. The stadium used to be the home of K.S.V. Waregem until the withdrawal of the club. It has 6,800 seats, with a full capacity of 8,500. The field is surrounded by a track. The Regenboogstadion does not meet UEFA's requirements for hosting UEFA Cup games, so Zulte Waregem's home games in the 2006-07 competition were played at the home of K.A.A. Gent. [1]

Honours

- **Belgian Second Division:**
  - **Winners (1):** 2004-05
- **Belgian Cup:**
  - **Winners (1):** 2005-06
- **Belgian Supercup:**
  - **Runners-up (1):** 2006

European record

As of December, 2008:

Season	Competition	Round	Nat.	Club	Home	Away
2006-07	UEFA Cup	1R		Lokomotiv Moskva	2-0	1-2
		Group F		Austria Wien		4-1
		Group F		Sparta Prague	3-1	
		Group F		Espanol Barcelona		2-6
		Group F		AFC Ajax	0-3	
		3R		Newcastle United	1-3	0-1

Current squad

As of 5 July, 2011

Note: Flags indicate national team as has been defined under FIFA eligibility rules. Players may hold more than one non-FIFA nationality.

No.	Position	Player
1	GK	Sammy Bossut
2	DF	Stijn Minne
3	DF	Steve Colpaert
4	DF	Davy De Fauw
6	MF	René Sterckx *(on loan from RSC Anderlecht)*
7	FW	Habib Habibou
8	DF	Brian Hamalainen
9	FW	Teddy Chevalier
10	FW	Aleksandar Trajkovski
12	MF	Thomas Matton
14	FW	Mbaye Leye
15	MF	*Niels Vandenbroucke*
16	MF	Ólafur Ingi Skúlason

No.	Position	Player
17	MF	Mikhail Sivakov
20	MF	*Thomas Van Der Haegen*
22	GK	Tom Muyters
23	DF	*Pavel Verstraete*
24	DF	Karel d'Haene
25	GK	*Jasper Otte*
26	FW	Jens Naessens
27	DF	Miguel Dachelet
28	MF	Jonathan Delaplace
29	MF	Jérémy Serwy
33	MF	Franck Berrier
	MF	Chahir Belghazouani
	MF	*Hernan Hinostroza*

Out on loan

Note: Flags indicate national team as has been defined under FIFA eligibility rules. Players may hold more than one non-FIFA nationality.

No.	Position	Player
14	FW	Jeremy Bokila *(to Sparta Rotterdam)*
19	FW	Giuseppe Rossini *(to Sint-Truiden)*

For recent transfers, see List of Belgian football transfers summer 2009.

Managers

- Francky Dury (2001-10)
- Bart De Roover (2010-Oct 10)
- Hugo Broos (Oct 2010-June 11)
- Darije Kalezić (Dec 2011)
- Francky Dury (2012-)

Notable former players

- Bart Buysse
- Nathan D'Haemers
- Frédéric Dupré
- Salou Ibrahim
- Pieter Merlier
- Tony Sergeant
- Rudison
- Nikica Jelavić
- Franck Berrier
- Jérémy Taravel
- Juan Diego González-Vigil
- Mbaye Leye
- Dragan Mrđa

References

[1] Roeder wary of falling flat against Belgian part-timers (http://football.guardian.co.uk/News_Story/0,,2013444,00.html)

External links

- (Dutch) Official Website (http://www.svzw.be)
- Zulte Waregem at UEFA.COM (http://www.uefa.com/footballeurope/club=89675/domestic.html)
- Zulte Waregem at EUFO.DE (http://www.eufo.de/football/bel/sv_zegem.htm)
- Zulte Waregem at Weltfusball.de (http://www.weltfussball.de/teams/sv-zulte-waregem/)
- Zulte Waregem at Playerhistory.com (http://www.playerhistory.com/club/1143/)
- Zulte Waregem at Transfermarkt.de (http://www.transfermarkt.de/de/verein/3508/sv-zulte-waregem/uebersicht/startseite.html)
- Zulte Waregem at Football Squads.co.uk (http://www.footballsquads.co.uk/belgium/2008-2009/eerste/zultewar.htm)
- Zulte Waregem at Football-Lineups.com (http://www.football-lineups.com/team/Zulte_Waregem/)

Arsenal_F.C.

Full name	Arsenal Football Club
Nickname(s)	The Gunners
Founded	1886 as *Dial Square*
Ground	Emirates Stadium, Holloway, London (Capacity: 60,361[1])
Owner	Arsenal Holdings plc
Chairman	Peter Hill-Wood
Manager	Arsène Wenger
League	Premier League
2010–11	Premier League, 4th
Website	Club home page [2]

Home colours **Away colours** **Third colours**

Arsenal Football Club is an English Premier League football club based in Holloway, London. One of the most successful clubs in English football, it has won 13 First Division and Premier League titles and 10 FA Cups. Arsenal holds the record for the longest uninterrupted period in the English top flight and would be placed first in an aggregated league of the entire 20th century.[3] It is the second side to complete an English top flight season unbeaten (in the 2003–04 season), and the only one to do it across 38 matches.

Arsenal was founded in 1886 in Woolwich and in 1893 became the first club from the south of England to join the Football League. In 1913, it moved north across the city to Arsenal Stadium in Highbury. In the 1930s the club won five League Championship titles and two FA Cups. After a lean period in the post-war years it won the League and FA Cup Double, in the 1970–71 season, and in the 1990s and first decade of the 21st century won two more Doubles and reached the 2006 UEFA Champions League Final. Arsenal has a long-standing rivalry with neighbours Tottenham Hotspur, with whom it contests the North London derby. Arsenal is the fourth most valuable association football club in the world as of 2012, valued at $1.3 billion.[4]

History

Woolwich Arsenal (in dark shirts) playing Newcastle United (in striped shirts) in an FA Cup semi-final — the club's first ever – at the Victoria Ground, Stoke.

Arsenal Football Club started out as Dial Square in 1886 by workers at the Royal Arsenal in Woolwich, south-east London, and was renamed Royal Arsenal shortly afterwards.[5] The club was renamed again to Woolwich Arsenal after becoming a limited company in 1893.[6] The club became the first southern member of the Football League in 1893, starting out in the Second Division, and won promotion to the First Division in 1904. The club's relative geographic isolation resulted in lower attendances than those of other clubs, which led to the club becoming mired in financial problems and effectively bankrupt by 1910, when they were taken over by businessmen Henry Norris and William Hall.[7] Norris sought to move the club elsewhere, and in 1913, soon after relegation back to the Second Division, Arsenal moved to the new Arsenal Stadium in Highbury, North London; they dropped "Woolwich" from their name the following year.[8] Arsenal only finished in fifth place in 1919, but were nevertheless elected to rejoin the First Division at the expense of local rivals Tottenham Hotspur, by reportedly dubious means.[9]

Arsenal appointed Herbert Chapman as manager in 1925. Having already won the league twice with Huddersfield Town in 1923–24 and 1924–25 (see Seasons in English football), Chapman brought Arsenal their first period of major success. His revolutionary tactics and training, along with the signings of star players such as Alex James and Cliff Bastin, laid the foundations of the club's domination of English football in the 1930s.[10] Under his guidance Arsenal won their first major trophies – victory in the 1930 FA Cup Final preceded two League Championships, in 1930–31 and 1932–33. In addition, Chapman was behind the 1932 renaming of the local London Underground station from "Gillespie Road" to "Arsenal", making it the only Tube station to be named specifically after a football club.[11]

Chapman died suddenly of pneumonia in early 1934, leaving Joe Shaw and George Allison to carry on his successful work. Under their guidance, Arsenal won three more titles, in 1933–34, 1934–35 and 1937–38, and the 1936 FA Cup. As key players retired, Arsenal had started to fade by the decade's end, and then the intervention of the Second World War meant competitive professional football in England was suspended.[12] [13] [14]

After the war, Arsenal enjoyed a second period of success under Allison's successor Tom Whittaker, winning the league in 1947–48 and 1952–53, and the FA Cup in 1950. Their fortunes waned thereafter; unable to attract players of the same calibre as they had in the 1930s, the club spent most of the 1950s and 1960s in trophyless mediocrity. Even former England captain Billy Wright could not bring the club any success as manager, in a stint between 1962 and 1966.[14] [15] [16]

Arsenal began winning silverware again with the surprise appointment of club physiotherapist Bertie Mee as manager in 1966. After losing two League Cup finals, they won their first European trophy, the 1969–70 Inter-Cities Fairs Cup. This was followed by an even greater triumph: their first League and FA Cup double in 1970–71.[17] This marked a premature high point of the decade; the Double-winning side was soon broken up and the following decade was characterised by a series of near misses. Arsenal finished as First Division runners-up in 1972–73, lost three FA Cup finals, in 1972, 1978 and 1980, and lost the 1980 Cup Winners' Cup final on penalties. The club's only success during this time was a last-minute 3–2 victory over Manchester United in the 1979 FA Cup Final, widely regarded as a classic.[14] [18]

The return of former player George Graham as manager in 1986 brought a third period of glory. Arsenal won the League Cup in 1986–87, Graham's first season in charge. This was followed by a League title win in 1988–89, won with a last-minute goal in the final game of the season against fellow title challengers Liverpool. Graham's Arsenal won another title in 1990–91, losing only one match, won the FA Cup and League Cup double in 1993, and a second

European trophy, the Cup Winners' Cup, in 1994.[14] [19] Graham's reputation was tarnished when he was found to have taken kickbacks from agent Rune Hauge for signing certain players,[20] and he was dismissed in 1995. His replacement, Bruce Rioch, lasted for only one season, leaving the club after a dispute with the board of directors.[21]

The club's success in the late 1990s and first decade of the 21st century owed a great deal to the 1996 appointment of Arsène Wenger as manager. Wenger brought new tactics, a new training regime and several foreign players who complemented the existing English talent. Arsenal won a second League and Cup double in 1997–98 and a third in 2001–02. In addition, the club reached the final of the 1999–2000 UEFA Cup (losing on penalties to Galatasaray), were victorious in the 2003 and 2005 FA Cups, and won the Premier League in 2003–04 without losing a single match, an achievement which earned the side the nickname "The Invincibles".[22] The club went 49 league matches unbeaten in that season, a national record.[23]

Arsenal's players and fans celebrate their 2004 League title win with an open-top bus parade.

Arsenal finished in either first or second place in the league in eight of Wenger's first eleven seasons at the club, although on no occasion were they able to retain the title.[14] As of March 2011, they were one of only four teams, the others being Manchester United, Blackburn Rovers and Chelsea, to have won the Premier League since its formation in 1992.[24] Arsenal had never progressed beyond the quarter-finals of the Champions League until 2005–06; in that season they became the first club from London in the competition's fifty-year history to reach the final, in which they were beaten 2–1 by Barcelona.[25] In July 2006, they moved into the Emirates Stadium, after 93 years at Highbury.[26]

Arsenal reached the final of the 2007 and 2011 League Cup final, losing 2–1 to Chelsea and Birmingham City respectively. The club has not gained a major trophy since the 2005 FA Cup.[27]

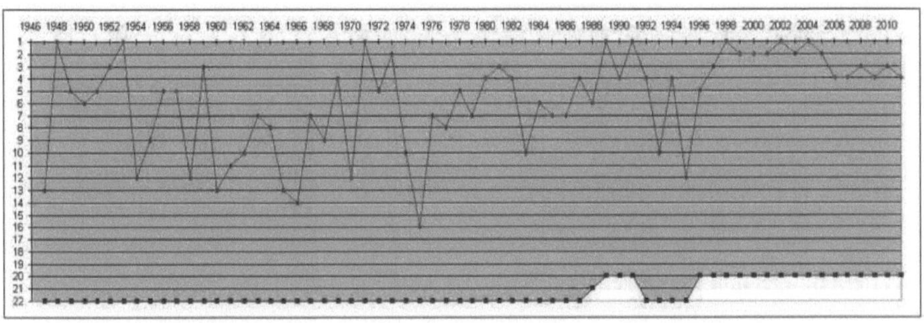

Arsenal's finishing positions since 1947 (in red). The number of teams in the top flight is shown in blue.

Crest

Unveiled in 1888, Royal Arsenal's first crest featured three cannon viewed from above, pointing northwards, similar to the coat of arms of the Metropolitan Borough of Woolwich. These can sometimes be mistaken for chimneys, but the presence of a carved lion's head and a cascabel on each are clear indicators that they are cannon.[28] This was dropped after the move to Highbury in 1913, only to be reinstated in 1922, when the club adopted a crest featuring a single cannon, pointing eastwards, with the club's nickname, *The Gunners*, inscribed alongside it; this crest only lasted until 1925, when the cannon was reversed to point westward and its barrel slimmed down.[28] In 1949, the club unveiled a modernised crest featuring the same style of cannon below the club's name, set in blackletter, and above the coat of arms of the Metropolitan Borough of Islington and a scroll inscribed with the club's newly-adopted Latin motto, *Victoria Concordia Crescit* "victory comes from harmony", coined by the club's programme editor Harry Homer.[28] For the first time, the crest was rendered in colour, which varied slightly over the crest's lifespan, finally becoming red, gold and green. Because of the numerous revisions of the crest, Arsenal were unable to copyright it. Although the club had managed to register the crest as a trademark, and had fought (and eventually won) a long legal battle with a local street trader who sold "unofficial" Arsenal merchandise,[29] Arsenal eventually sought a more comprehensive legal protection. Therefore, in 2002 they introduced a new crest featuring more modern curved lines and a simplified style, which was copyrightable.[30] The cannon once again faces east and the club's name is written in a sans-serif typeface above the cannon. Green was replaced by dark blue. The new crest was criticised by some supporters; the Arsenal Independent Supporters' Association claimed that the club had ignored much of Arsenal's history and tradition with such a radical modern design, and that fans had not been properly consulted on the issue.[31]

Until the 1960s, a badge was worn on the playing shirt only for high-profile matches such as FA Cup finals, usually in the form of a monogram of the club's initials in red on a white background.[32]

The monogram theme was developed into an Art Deco-style badge on which the letters A and C framed a football rather than the letter F, the whole set within a hexagonal border. This early example of a corporate logo, introduced as part of Herbert Chapman's rebranding of the club in the 1930s, was used not only on Cup Final shirts but as a design feature throughout Highbury Stadium, including above the main entrance and inlaid in the floors.[33] From 1967, a white cannon was regularly worn on the shirts, until replaced by the club crest, sometimes with the addition of the nickname "The Gunners", in the 1990s.[32]

In the 2011–2012 season, Arsenal will celebrate their 125th year anniversary. The celebrations will include a modified version of the current crest worn on their jerseys for the season. The crest is all white, surrounded by 15 oak leaves to the right and 15 laurel leaves to the left. The oak leaves represent the 15 founding members of the club who met at the Royal Oak pub. The 15 laurel leaves represent the design detail on the six pence pieces paid by the founding fathers to establish the club. The laurel leaves also represent strength.To complete the crest, 1886 and 2011 are shown on either sides of the motto "Forward" at the bottom of the crest.[34]

Arsenal's first crest from 1888

The 'monogram' badge as used in the 1930 FA Cup Final

The 'Art Deco' badge as used in the 1952 FA Cup Final

A version of the Arsenal crest used from 1949 to 2002

The Arsenal logo used from 2002 onwards (with the exception of the 2011–12 season)

The Arsenal logo, which commemorated its 125th anniversary, was used only in the 2011–12 season

Colours

Arsenal's original home colours. The team wore a similar kit (but with redcurrant socks) during the 2005–06 season.

For much of Arsenal's history, their home colours have been bright red shirts with white sleeves and white shorts, though this has not always been the case. The choice of red is in recognition of a charitable donation from Nottingham Forest, soon after Arsenal's foundation in 1886. Two of Dial Square's founding members, Fred Beardsley and Morris Bates, were former Forest players who had moved to Woolwich for work. As they put together the first team in the area, no kit could be found, so Beardsley and Bates wrote home for help and received a set of kit and a ball.[5] The shirt was redcurrant, a dark shade of red, and was worn with white shorts and blue socks.[35]

In 1933, Herbert Chapman, wanting his players to be more distinctly dressed, updated the kit, adding white sleeves and changing the shade to a brighter pillar box red. Two possibilities have been suggested for the origin of the white sleeves. One story reports that Chapman noticed a supporter in the stands wearing a red sleeveless sweater over a white shirt; another was that he was inspired by a similar outfit worn by the cartoonist Tom Webster, with whom Chapman played golf.[36] Regardless of which story is true, the red and white shirts have come to define Arsenal and the team have worn the combination ever since, aside from two seasons. The first was 1966–67, when Arsenal wore all-red shirts;[35] this proved unpopular and the white sleeves returned the following season. The second was 2005–06, the last season that Arsenal played at Highbury, when the team wore commemorative redcurrant shirts similar to those worn in 1913, their first season in the stadium; the club reverted to their normal colours at the start of the next season.[36] In the 2008–09 season, Arsenal replaced the traditional all-white sleeves with red sleeves with a broad white stripe.[35]

Arsenal's home colours have been the inspiration for at least three other clubs. In 1909, Sparta Prague adopted a dark red kit like the one Arsenal wore at the time;[36] in 1938, Hibernian adopted the design of the Arsenal shirt sleeves in their own green and white strip.[37] In 1920, Sporting Clube de Braga's manager returned from a game at Highbury and changed his team's green kit to a duplicate of Arsenal's red with white sleeves and shorts, giving rise to the team's nickname of *Os Arsenalistas*.[38] These teams still wear these designs to this day.

For many years Arsenal's away colours were white shirts and either black or white shorts. Since the 1969–70 season, they have worn yellow and blue, but there have been exceptions. They wore a green and navy away kit in 1982–83, and since the early 1990s and the advent of the lucrative replica kit market, the away colours have been changed regularly. During this period the designs have been either two-tone blue designs, or variations on the traditional yellow and blue, such as the metallic gold and navy strip used in the 2001–02 season, and the yellow and dark grey used from 2005 to 2007.[39] As of 2009, the away kit is changed every season, and the outgoing away kit becomes the third-choice kit if a new home kit is being introduced in the same year.[40]

Arsenal's shirts have been made by manufacturers including Bukta (from the 1930s until the early 1970s), Umbro (from the 1970s until 1986), Adidas (1986–1994), and Nike (since 1994). Like those of most other major football clubs, Arsenal's shirts have featured sponsors' logos since the 1980s; sponsors include JVC (1982–1999), Sega (1999–2002), O2 (2002–2006), and Emirates (from 2006).[35] [36]

Stadiums

The North Bank Stand, Arsenal Stadium, Highbury

For most of their time in south-east London, Arsenal played at the Manor Ground in Plumstead, apart from a three-year period at the nearby Invicta Ground between 1890 and 1893. The Manor Ground was initially just a field, until the club installed stands and terracing for their first Football League match in September 1893. They played their home games there for the next twenty years (with two exceptions in the 1894–95 season), until the move to north London in 1913.[41] [42]

Widely referred to as Highbury, Arsenal Stadium was the club's home from September 1913 until May 2006. The original stadium was designed by the renowned football architect Archibald Leitch, and had a design common to many football grounds in the UK at the time, with a single covered stand and three open-air banks of terracing.[43] The entire stadium was given a massive overhaul in the 1930s: new Art Deco West and East stands were constructed, opening in 1932 and 1936 respectively, and a roof was added to the North Bank terrace, which was bombed during the Second World War and not restored until 1954.[43]

Highbury could hold over 60,000 spectators at its peak, and had a capacity of 57,000 until the early 1990s. The Taylor Report and Premier League regulations obliged Arsenal to convert Highbury to an all-seater stadium in time for the 1993–94 season, thus reducing the capacity to 38,419 seated spectators.[44] This capacity had to be reduced further during Champions League matches to accommodate additional advertising boards, so much so that for two seasons, from 1998 to 2000, Arsenal played Champions League home matches at Wembley, which could house more than 70,000 spectators.[45]

The Emirates Stadium filling up on the day of Dennis Bergkamp's testimonial

Expansion of Highbury was restricted because the East Stand had been designated as a Grade II listed building and the other three stands were close to residential properties.[43] These limitations prevented the club from maximising matchday revenue during the 1990s and first decade of the 21st century, putting them in danger of being left behind in the football boom of that time.[46] After considering various options, in 2000 Arsenal proposed building a new 60,361-capacity stadium at Ashburton Grove, since renamed the Emirates Stadium, about 500 metres south-west of Highbury.[47] The project was initially delayed by red tape and rising costs,[48] and construction was completed in July 2006, in time for the start of the 2006–07 season.[49] The stadium was named after its sponsors, the airline company Emirates, with whom the club signed the largest sponsorship deal in English football history, worth around £100 million;[50] some fans referred to the ground as Ashburton Grove, or the Grove, as they did not agree with corporate sponsorship of stadium names.[51] The stadium will be officially known as Emirates Stadium until at least 2012, and the airline will be the club's shirt sponsor until the end of the 2013–14 season.[50] From the start of the 2010–11 season on, the stands of the stadium have been officially known as North Bank, East Stand, West Stand and Clock end.[52]

Arsenal's players train at the Shenley Training Centre in Hertfordshire, a purpose-built facility which opened in 1999.[53] Before that the club used facilities on a nearby site owned by the University College of London Students' Union. Until 1961 they had trained at Highbury.[54] Arsenal's Academy under-18 teams play their home matches at Shenley, while the reserves play their games at Underhill, home of Barnet.[55]

Supporters

Arsenal against rivals Tottenham, known as the North London derby, in November 2010.

Arsenal fans often refer to themselves as "Gooners", the name derived from the team's nickname, "The Gunners". The fanbase is large and generally loyal, and virtually all home matches sell out; in 2007–08 Arsenal had the second-highest average League attendance for an English club (60,070, which was 99.5% of available capacity),[56] and as of 2006, the fourth-highest all-time average attendance.[57] Arsenal has the seventh highest average attendance of European football clubs only behind Borussia Dortmund, FC Barcelona, Manchester United, Real Madrid, Bayern Munich, and Schalke.[58] [59] [60] [61] The club's location, adjoining wealthy areas such as Canonbury and Barnsbury, mixed areas such as Islington, Holloway, Highbury, and the adjacent London Borough of Camden, and largely working-class areas such as Finsbury Park and Stoke Newington, has meant that Arsenal's supporters have come from across the usual class divides.

Like all major English football clubs, Arsenal have a number of domestic supporters' clubs, including the Arsenal Football Supporters Club, which works closely with the club, and the Arsenal Independent Supporters' Association, which maintains a more independent line. The Arsenal Supporters' Trust promotes greater participation in ownership of the club by fans. The club's supporters also publish fanzines such as *The Gooner*, *Highbury High*, *Gunflash* and the less cerebral *Up The Arse!*. In addition to the usual English football chants, supporters sing "One-Nil to the Arsenal" (to the tune of "Go West") and "Boring, Boring Arsenal", which used to be a common taunt from opposition fans but is now sung ironically by Arsenal supporters when the team is playing well.[62]

There have always been Arsenal supporters outside of London, and since the advent of satellite television, a supporter's attachment to a football club has become less dependent on geography. Consequently, Arsenal have a significant number of fans from beyond London and all over the world; in 2007, 24 UK, 37 Irish and 49 other overseas supporters clubs were affiliated with the club.[63] A 2005 report by Granada Ventures, which at the time owned a 9.9% stake in the club, estimated Arsenal's global fanbase at 27 million.[64]

Arsenal's longest-running and deepest rivalry is with their nearest major neighbours, Tottenham Hotspur; matches between the two are referred to as North London derbies.[65] Other rivalries within London include those with Chelsea, Fulham and West Ham United. In addition, Arsenal and Manchester United developed a strong on-pitch rivalry in the late 1980s, which intensified in recent years when both clubs were competing for the Premier League title[66] – so much so that a 2003 online poll by the Football Fans Census listed Manchester United as Arsenal's biggest rivals, followed by Tottenham and Chelsea.[67] A 2008 poll listed the Tottenham rivalry as more important.[68]

Ownership and finances

Arsenal's parent company, Arsenal Holdings plc, operates as a non-quoted public limited company, whose ownership is considerably different from that of other football clubs. Only 62,217 shares in Arsenal have been issued,[69] [70] and they are not traded on a public exchange such as the FTSE or AIM; instead, they are traded relatively infrequently on PLUS (AFC [71]), a specialist market. At 31 August 2010, a single share in Arsenal had a mid price of £10,250, which set the club's market capitalisation value at approximately £637.74m.[72] The club made a pre-tax operating profit (excluding player transfers) of £62.7m in the year ending 31 May 2009, from a turnover of £313.3m.[73]

The largest shareholder on the Arsenal board is American sports tycoon Stan Kroenke, who launched a bid for the club in 2007,[74] and in November 2009 increased his holding to 18,594 shares (29.9%).[75]

A rival bid to Kroenke's came from Red & White Securities, which is co-owned by Russian billionaire Alisher Usmanov and London-based financier Farhad Moshiri.[76] Red & White launched its bid in August 2007, buying the stake held by former Arsenal vice-chairman David Dein, and as at February 2009 owned 15,555 shares (25.0%) in the club.[77] This led to press speculation of a bidding war between Kroenke and Usmanov.[76] However, Kroenke agreed not to purchase more than 29.9% of the club until at least September 2009,[78] while the rest of the board have first option on each others' shares until October 2012.[79]

In April 2011, Stan Kroenke was rumoured to be in "advanced talks" with Arsenal over a possible takeover.[80]

As of October 2011, Kroenke owns 41,574 shares (66.82%) and Red & White Securities own 18,261 shares (29.35%).[81] Under company law Kroenke, as majority shareholder, is obliged to make an offer for the remaining shares in the club.

Kit manufacturers and shirt sponsors

Period	Kit manufacturer	Shirt sponsor
1930s–1970	Bukta	*none*
1971–1980	Umbro	
1981–1986		JVC
1986–1994	Adidas	
1994–1999	Nike	
1999–2002		Dreamcast / Sega[1]
2002–2006		O2
2006–		Fly Emirates

[1] Dreamcast was worn on the first and third shirt while Sega was printed on the second shirt.

In popular culture

Arsenal have appeared in a number of media "firsts". On 22 January 1927, their match at Highbury against Sheffield United was the first English League match to be broadcast live on radio.;[82] [83] A decade later, on 16 September 1937, an exhibition match between Arsenal's first team and the reserves was the first football match in the world to be televised live.[82] [84] Arsenal also featured in the first edition of the BBC's *Match of the Day*, which screened highlights of their match against Liverpool at Anfield on 22 August 1964.[82] [85] BSkyB's coverage of Arsenal's January 2010 match against Manchester United was the first live public broadcast of a sports event on 3D television.[82] [86]

As one of the most successful teams in the country, Arsenal have often featured when football is depicted in the arts in Britain. They formed the backdrop to one of the earliest football-related films, *The Arsenal Stadium Mystery* (1939).[87] The film centres on a friendly match between Arsenal and an amateur side, one of whose players is poisoned while playing. Many Arsenal players appeared as themselves and manager George Allison was given a speaking part.[88] More recently, the book *Fever Pitch* by Nick Hornby was an autobiographical account of Hornby's life and relationship with football and Arsenal in particular. Published in 1992, it formed part of the revival and rehabilitation of football in British society during the 1990s.[89] The book was twice adapted for the cinema – the 1997 British film focuses on Arsenal's 1988–89 title win,[90] and a 2005 American version features a fan of baseball's Boston Red Sox.[91]

Arsenal have often been stereotyped as a defensive and "boring" side, especially during the 1970s and 1980s;[62] [92] many comedians, such as Eric Morecambe, made jokes about this at the team's expense. The theme was repeated in the 1997 film *The Full Monty*, in a scene where the lead actors move in a line and raise their hands, deliberately mimicking the Arsenal defence's offside trap, in an attempt to co-ordinate their striptease routine.[88] Another film reference to the club's defence comes in the film *Plunkett & Macleane*, in which two characters are named Dixon and Winterburn after Arsenal's long-serving full backs – the right-sided Lee Dixon and the left-sided Nigel Winterburn.[88]

The 1991 television comedy sketch show *Harry Enfield & Chums* featured a sketch from the characters Mr Cholmondly-Warner and Grayson where the Arsenal team of 1933, featuring exaggerated parodies of fictitious amateur players take on the Liverpool team of 1991.[93]

Arsenal ladies

Arsenal Ladies is the women's football club affiliated to Arsenal. Founded in 1987, they turned semi-professional in 2002 and are managed by Laura Harvey.[94] Arsenal Ladies are the most successful team in English women's football. In the 2008–09 season, they won all three major English trophies – the FA Women's Premier League, FA Women's Cup and FA Women's Premier League Cup,[95] and, as of 2009, were the only English side to have won the UEFA Women's Cup, having done so in the 2006–07 season as part of a unique quadruple.[96] The men's and women's clubs are formally separate entities but have quite close ties; Arsenal Ladies are entitled to play once a season at the Emirates Stadium, though they usually play their home matches at Boreham Wood.[97]

In the community

In 1985, Arsenal founded a community scheme, "Arsenal in the Community", which offered sporting, social inclusion, educational and charitable projects. The club support a number of charitable causes directly and in 1992 established The Arsenal Charitable Trust, which by 2006 had raised more than £2 million for local causes.[98] An ex-professional and celebrity football team associated with the club also raised money by playing charity matches.[99]

In the 2009/10 season Arsenal announced that they had raised a record breaking £818,897 for the Great Ormond Street Hospital Children's Charity. The original target was £500,000 but thanks to the overwhelming support from fans, players, directors and staff. They were able to smash the target.[100]

On September 2011, Arsenal won their case against Alicia Simon, a Seville resident who has now been forced by the Spanish Patent and Trademark Office to change the name of her shop "Arsenale". Ms Simon registered the name of her shop before she even opened it in 2007 and has faced litigation from the North London club ever since.[101]

Statistics and records

David O'Leary holds the record for Arsenal appearances, having played 722 first-team matches between 1975 and 1993. Fellow centre half and former captain Tony Adams comes second, having played 669 times. The record for a goalkeeper is held by David Seaman, with 564 appearances.[102]

Thierry Henry is the club's top goalscorer with 229 goals in all competitions between 1999 and 2012,[103] having surpassed Ian Wright's total of 185 in October 2005.[104] Wright's record had stood since September 1997, when he overtook the longstanding total of 178 goals set by winger Cliff Bastin in 1939.[105] Henry also holds the club record for goals scored in the League, with 176,[103] a record that had been held by Bastin until February 2006.[106]

Arsenal's record home attendance is 73,707, for a UEFA Champions League match against RC Lens on 25 November 1998 at Wembley Stadium, where the club formerly played home European matches because of the limits on Highbury's capacity. The record attendance for an Arsenal match at Highbury is 73,295, for a 0–0 draw against Sunderland on 9 March 1935,[102] while that at Emirates Stadium is 60,161, for a 2–2 draw with Manchester United on 3 November 2007.[107]

Arsenal have also set records in English football, including the most consecutive seasons spent in the top flight (84 as of 2010–11) and the longest run of unbeaten League matches (49 between May 2003 and October 2004).[23] This included all 38 matches of their title-winning 2003–04 season, when Arsenal became only the second club to finish a top-flight campaign unbeaten, after Preston North End (who played only 22 matches) in 1888–89.[22]

Arsenal also set a Champions League record during the 2005–06 season by going ten matches without conceding a goal, beating the previous best of seven set by A.C. Milan. They went a record total stretch of 995 minutes without letting an opponent score; the streak ended in the final, when Samuel Eto'o scored a 76th-minute equaliser for Barcelona.[25]

Players

First-team squad

As of 18 February 2012.[108]

Note: Flags indicate national team as has been defined under FIFA eligibility rules. Players may hold more than one non-FIFA nationality.

No.	Position	Player
1	GK	Manuel Almunia
2	MF	Abou Diaby
3	DF	Bacary Sagna
4	DF	Per Mertesacker
5	DF	Thomas Vermaelen *(vice-captain)*
6	DF	Laurent Koscielny
7	MF	Tomáš Rosický
8	MF	Mikel Arteta
9	FW	Park Chu-Young
10	FW	Robin van Persie *(captain)*
11	DF	André Santos
13	GK	Wojciech Szczęsny
14	FW	Theo Walcott
15	FW	Alex Oxlade-Chamberlain

No.	Position	Player
16	MF	Aaron Ramsey
17	MF	Alex Song
18	DF	Sébastien Squillaci
19	MF	Jack Wilshere
20	DF	Johan Djourou
21	GK	Łukasz Fabiański
25	DF	Carl Jenkinson
26	MF	Emmanuel Frimpong
27	FW	Gervinho

No.		Position	Player
28		DF	Kieran Gibbs
29		FW	Marouane Chamakh
30		MF	Yossi Benayoun *(on loan from Chelsea)*
39		MF	Francis Coquelin

For recent transfers, see 2011–12 Arsenal F.C. season.

Players signing at end of current season:

Note: Flags indicate national team as has been defined under FIFA eligibility rules. Players may hold more than one non-FIFA nationality.

No.	Position	Player
	FW	Lukas Podolski

Reserve squad

As of 17 March 2012.[109] [110]

Note: Flags indicate national team as has been defined under FIFA eligibility rules. Players may hold more than one non-FIFA nationality.

No.	Position	Player
35	FW	Zak Ansah
37	DF	Daniel Boateng
38	DF	George Brislen-Hall
44	MF	Conor Henderson
45	DF	Gavin Hoyte
47	GK	Sean McDermott

No.	Position	Player
48	MF	Jernade Meade
49	DF	Ignasi Miquel
51	FW	Nigel Neita
53	MF	Oğuzhan Özyakup
61	MF	Josh Rees
62	GK	James Shea

Out on loan

Note: Flags indicate national team as has been defined under FIFA eligibility rules. Players may hold more than one non-FIFA nationality.

No.	Position	Player
23	MF	Andrei Arshavin *(at Zenit Saint Petersburg until the end of the 2011–12 season)*[111]
24	GK	Vito Mannone *(at Hull City until the end of the 2011–12 season)*[112]
31	FW	Ryo Miyaichi *(at Bolton Wanderers until the end of the 2011–12 season)*[113]
33	FW	Benik Afobe *(at Reading until the end of the 2011–12 season)*[114]
34	MF	Chuks Aneke *(at Preston North End until the end of the 2011-12 season)*[115]
40	MF	Craig Eastmond *(at Wycombe Wanderers until the end of the 2011–12 season)*[116]
43	DF	Sead Hajrović *(at Barnet until the end of the 2011–12 season)*[117]
46	MF	Henri Lansbury *(at West Ham United until the end of the 2011–12 season)*[118]

No.	Position	Player
50	FW	Rhys Murphy *(at Preston North End until the end of the 2011–12 season)*[119]
52	FW	Nicklas Bendtner *(at Sunderland until the end of the 2011–12 season)*[120]
55	MF	Sanchez Watt *(at Crawley Town until the end of the 2011–12 season)*[121]
56	MF	Nico Yennaris *(at Notts County until the end of the 2011–12 season)*[122]
—	DF	Kyle Bartley *(at Rangers until the end of the 2011–12 season)*[123]
—	FW	Joel Campbell *(at Lorient until the end of the 2011–12 season)*[124]
—	MF	Denílson *(at São Paulo until the end of the 2011–12 season)*[125]
—	FW	Carlos Vela *(at Real Sociedad until the end of the 2011–12 season)*[126]

Current technical staff

As of February 2012.[127] [128] [129]

Arsène Wenger, who has been managing Arsenal since 1996.

Position	Name
Manager	Arsène Wenger
Assistant manager	Pat Rice
First team coach	Boro Primorac
Reserve team coach	Neil Banfield
Youth team coach	Steve Bould
Goalkeeping coach	Gerry Peyton
Fitness coach	Tony Colbert
Physiotherapist	Colin Lewin
Assistant physiotherapists	David Wales Simon Harland
Club doctor	Gary O'Driscoll
Kit manager	Vic Akers
Chief scout	Steve Rowley
Head of youth development	Liam Brady

Managers

There have been eighteen permanent and five caretaker managers of Arsenal since the appointment of the club's first professional manager, Thomas Mitchell in 1897.[130] The club's longest-serving manager as of 2009, in terms of both length of tenure and number of games overseen, is Arsène Wenger, who was appointed in 1996.[131] [132] Wenger is also Arsenal's only manager from outside the United Kingdom.[132] Two Arsenal managers have died in the job – Herbert Chapman and Tom Whittaker.[133]

Honours

Domestic

- **First Division** (until 1992) and **Premier League**[14] [134]

13: 1930–31, 1932–33, 1933–34, 1934–35, 1937–38, 1947–48, 1952–53, 1970–71, 1988–89, 1990–91, 1997–98, 2001–02, 2003–2004

- **FA Cup**[14]

10: 1930, 1936, 1950, 1971, 1979, 1993, 1998, 2002, 2003, 2005

- **League Cup**[14]

2: 1987, 1993

- **FA Community Shield** (**FA Charity Shield** before 2002)[135]

12: 1930, 1931, 1933, 1934, 1938, 1948, 1953, 1991 (shared), 1998, 1999, 2002, 2004

European

- **European Cup Winners' Cup**[14]

1: 1993–94

- **Inter-Cities Fairs Cup**[14]

1: 1969–70

Notable statistics

Arsenal's tally of 13 League Championships is the third highest in English football, after Manchester United (19) and Liverpool (18),[136] while the total of 10 FA Cups is the second highest, after Manchester United (11).[137] Arsenal have achieved three League and FA Cup "Doubles" (in 1971, 1998 and 2002), a feat only previously achieved by Manchester United (in 1994, 1996 and 1999),[14] [138] and in 1993 were the first side in English football to complete the FA Cup and League Cup double.[139] They were also the first London club to reach the final of the UEFA Champions League, in 2006, losing the final 2–1 to Barcelona.[140]

Arsenal have one of the best top-flight records in history, having finished below fourteenth only seven times. Arsenal also have the highest average league finishing position for the period 1900–1999, with an average league placing of 8.5.[3] In addition, they are one of only six clubs to have won the FA Cup twice in succession, in 2002 and 2003.[141] Arsenal also hold the record for the longest unbeaten run in the Premier League at 49 games, and are the only team to have gone an entire Premier League season unbeaten in 2003–04.[142]

Footnotes

[1] http://www.uefa.com/MultimediaFiles/Download/StatDoc/competitions/UCL/01/67/63/78/1676378_DOWNLOAD.pdf

[2] http://www.arsenal.com/

[3] "Football: How consistency and caution made Arsenal England's greatest team of the 20th century" (http://www.independent.co.uk/sport/football-how-consistency-and-caution-made-arsenal-englands-greatest-team-of-the-20th-century-1133020.html). The Independent. . Retrieved 27 April 2012.

[4] "Arsenal" (http://www.forbes.com/sites/mikeozanian/2012/04/18/manchester-united-again-the-worlds-most-valuable-soccer-team/). *Forbes*. 18 April 2012. . Retrieved 22 April 2012.

[5] Soar, Phil & Tyler, Martin (2005). *The Official Illustrated History of Arsenal*. Hamlyn. p. 23. ISBN 978-0-600-61344-2.

[6] "Royal Arsenal becomes Woolwich Arsenal" (http://homepage.ntlworld.com/andy.kelly/arsenal/record/namechange.htm). *Andy Kelly's Arsenal Resource Website*. . Retrieved 13 October 2010.

[7] Soar & Tyler (2005). *The Official Illustrated History of Arsenal*. pp. 32–33.

[8] Soar & Tyler (2005). *The Official Illustrated History of Arsenal*. p. 40.

[9] It has been alleged that Arsenal's promotion, on historical grounds rather than merit, was thanks to underhand actions by Norris, who was chairman of the club at the time; see History of Arsenal F.C. (1886–1966) for more details. These allegations range from political machinations to outright bribery; no firm proof of any wrongdoing has ever been found. A brief account is given in Soar & Tyler (2005). *The Official Illustrated History of Arsenal*. p. 40. A more detailed account can be found in Spurling, Jon (2004). *Rebels for the Cause: The Alternative History of Arsenal Football Club*. Mainstream. pp. 38–41. ISBN 978-1-84018-900-1.

[10] Soar & Tyler (2005). *The Official Illustrated History of Arsenal*. p. 18.

[11] "London Underground and Arsenal present The Final Salute to Highbury" (http://www.tfl.gov.uk/static/corporate/media/newscentre/archive/3709.html). Transport for London. 12 January 2006. . Retrieved 12 August 2011.

[12] "Arsenal clinch a hat-trick of titles" (http://www.arsenal.com/history/herbert-chapman/arsenal-clinch-a-hat-trick-of-titles). Arsenal F.C. . Retrieved 27 November 2009.

[13] Brown, Tony (2007). *Champions all!* (http://www.soccer.mistral.co.uk/books/ch6-10.pdf). Nottingham: SoccerData. pp. 6–7. ISBN 1-905891-02-4. .

[14] "Arsenal" (http://www.fchd.info/ARSENAL.HTM). *Football Club History Database*. Richard Rundle. . Retrieved 23 October 2009.

[15] "Post-War Arsenal – Overview" (http://www.arsenal.com/history/post-war-arsenal/post-war-arsenal-overview). Arsenal F.C. . Retrieved 27 November 2009.

[16] Brown (2007). *Champions all!*. p. 7.

[17] Galvin, Robert. "Bertie Mee" (http://www.nationalfootballmuseum.com/pages/fame/Inductees/bertiemee.htm). *Football Hall of Fame*. National Football Museum. . Retrieved 23 October 2009.

[18] A 2005 poll of English football fans rated the 1979 FA Cup Final the 15th greatest game of all time. Reference: Winter, Henry (19 April 2005). "Classic final? More like a classic five minutes" (http://www.telegraph.co.uk/sport/main.jhtml?xml=/sport/2005/04/19/sfnwin19.xml). *Daily Telegraph*. . Retrieved 11 August 2008.

[19] Clarke, Andy (26 March 2009). "Top Ten: Title Run-ins" (http://www.skysports.com/story/0,19528,11670_3335042,00.html). Sky Sports. . Retrieved 7 December 2009.

[20] Graham was banned for a year by the Football Association for his involvement in the scandal after he admitted he had received an "unsolicited gift" from Hauge. Reference: Collins, Roy (18 March 2000). "Rune Hauge, international man of mystery" (http://football.guardian.co.uk/News_Story/0,,148114,00.html). *The Guardian*. . Retrieved 11 August 2008. The case is given a detailed treatment in Bower, Tom (2003). *Broken Dreams*. Simon & Schuster. ISBN 978-0-7434-4033-2.

[21] Moore, Glenn (13 August 1996). "Rioch at odds with the system" (http://www.independent.co.uk/sport/rioch-at-odds-with-the-system-1309564.html). *The Independent*. . Retrieved 23 October 2009.

[22] Hughes, Ian (15 May 2004). "Arsenal the Invincibles" (http://news.bbc.co.uk/sport1/hi/football/teams/a/arsenal/3713537.stm). BBC Sport. . Retrieved 11 August 2008.

[23] Fraser, Andrew (25 October 2004). "Arsenal run ends at 49" (http://news.bbc.co.uk/sport1/hi/football/teams/a/arsenal/3950453.stm). BBC Sport. . Retrieved 27 August 2008.

[24] Ross, James M. "FA Premier League Champions 1993–2007" (http://www.rsssf.com/tablese/engchamp.html#c1993). Rec.Sport.Soccer Statistics Foundation. . Retrieved 11 August 2008.

[25] "2005/06: Ronaldinho delivers for Barça" (http://www.uefa.com/competitions/ucl/history/season=2005/intro.html). UEFA. 17 May 2007. . Retrieved 11 August 2008.

[26] Aizlewood, John (23 July 2006). "Farewell Bergkamp, hello future" (http://www.timesonline.co.uk/tol/sport/football/article691484.ece). *The Times* (UK). . Retrieved 23 October 2009.

[27] Gilbert, Mark (15 September 2011). "Cesc Fabregas: Arsenal Title Drought is Far From Over" (http://www.thesun.co.uk/sol/homepage/sport/football/3816819/Cesc-Fabregas-Arsenal-trophy-drought-is-far-from-over.htmll). *The Sun* (UK). . Retrieved 15 September 2011.

[28] "The Crest" (http://www.arsenal.com/history/the-crest). Arsenal F.C. . Retrieved 11 August 2008.

[29] Free, Dominic (4 June 2003). "*Arsenal v. Reed* in the Court of Appeal" (http://www.simkins.co.uk/ebulletins/DAFArsenalReed.aspx). Michael Simkins LLP. . Retrieved 11 August 2008.

[30] "Arsenal go for a makeover" (http://news.bbc.co.uk/sport1/hi/football/teams/a/arsenal/1795444.stm). BBC Sport. 1 February 2004. . Retrieved 11 August 2008.

[31] "Crestfallen" (http://web.archive.org/web/20061108191039/http://www.aisa.org/pdfs/crest_leaflet.pdf) (PDF). Arsenal Independent Supporters' Association. Archived from the original (http://www.aisa.org/pdfs/crest_leaflet.pdf) on 8 November 2006. . Retrieved 11 August 2008.

[32] "The Arsenal shirt badge" (http://www.arsenal.com/news/news-archive/the-arsenal-shirt-badge). Arsenal F.C. . Retrieved 24 January 2010.

[33] "The Art Deco crest" (http://www.arsenal.com/news/news-archive/the-art-deco-crest). Arsenal F.C. . Retrieved 24 January 2010.

[34] "125th anniversary crest" (http://www.arsenal.com/news/news-archive/125-years-in-the-making-anniversary-crest). Arsenal F.C. . Retrieved 1 May 2011.

[35] "Arsenal" (http://www.historicalkits.co.uk/Arsenal/Arsenal.htm). *Historical Football Kits*. D & M Moor. . Retrieved 8 December 2006.

[36] "Arsenal Kit Design" (http://www.arsenal.com/history/kit-design). Arsenal F.C. . Retrieved 8 November 2008.

[37] "Hibernian" (http://www.historicalkits.co.uk/Scottish_Football_League/Hibernian/hibernian.htm). *Historical Football Kits*. D & M Moor. . Retrieved 11 August 2008.

[38] Rui Matos Pereira (21 October 2005). "Secret of Braga's success" (https://video.uefa.com/news/newsid=356684.html). UEFA. . Retrieved 23 January 2011.

[39] "Arsenal Change Kits" (http://www.historicalkits.co.uk/Arsenal/Arsenal-change-kits.html). *Historical Football Kits*. D & M Moor. . Retrieved 27 November 2009.

[40] "Club Charter" (http://www.arsenal.com/the-club/corporate-info/the-club-charter). Arsenal F.C. . Retrieved 23 October 2009.

[41] Inglis, Simon (1996) [1985]. *Football Grounds of Britain* (3rd ed.). London: CollinsWillow. pp. 16–17. ISBN 0-00-218426-5.

[42] "Suspension of the Plumstead Ground". *The Times*: p. 6. 7 February 1895.

[43] "A Conservation Plan for Highbury Stadium, London" (http://www.islington.gov.uk/DownloadableDocuments/Environment/Pdf/highburyconservationplan_2005.pdf) (PDF). Islington Council. February 2005. . Retrieved 11 August 2008.

[44] "Highbury" (http://web.archive.org/web/20080111223345/http://www.arsenal.com/article.asp?thisNav=The+Club&article=344883&Title=Highbury). Arsenal F.C. Archived from the original (http://www.arsenal.com/article.asp?thisNav=The+Club&article=344883&Title=Highbury) on 11 January 2008. .

[45] "Arsenal get Wembley go-ahead" (http://news.bbc.co.uk/1/hi/sport/football/138932.stm). BBC Sport. 24 July 1998. . Retrieved 11 August 2008.

[46] Garner, Clare (18 August 1997). "Arsenal consider leaving hallowed marble halls" (http://www.independent.co.uk/sport/arsenal-consider-leaving-hallowed-marble-halls-1246012.html). *The Independent*. . Retrieved 23 October 2009.

[47] "Arsenal unveil new stadium plans" (http://news.bbc.co.uk/sport1/hi/football/teams/a/arsenal/1011234.stm). BBC Sport. 7 November 2000. . Retrieved 11 August 2008.

[48] "Arsenal stadium delay" (http://news.bbc.co.uk/sport1/hi/football/teams/a/arsenal/2953273.stm). BBC Sport. 16 April 2003. . Retrieved 11 August 2008.

[49] "Bergkamp given rousing farewell" (http://news.bbc.co.uk/sport2/hi/football/teams/a/arsenal/5203954.stm). BBC Sport. 22 July 2006. . Retrieved 23 August 2007.

[50] "Arsenal name new ground" (http://news.bbc.co.uk/sport1/hi/football/teams/a/arsenal/3715678.stm). BBC Sport. 5 October 2004. . Retrieved 11 August 2008.

[51] Dawes, Brian (2006). "The 'E' Word" (http://www.arsenal-world.co.uk/news/loadnews.asp?cid=TMNW&id=283908). *Arsenal World*. Footymad. . Retrieved 11 August 2008.

[52] Emirates Stadium stands to be re-named (http://www.arsenal.com/news/news-archive/emirates-stadium-stands-to-be-re-named) Arsenal FC, 19 July 2010

[53] Taylor, David (21 October 1999). "Arsenal gets a complex" (http://www.architectsjournal.co.uk/arsenal-gets-a-complex/774448.article). *The Architects' Journal*. . Retrieved 20 January 2010.

[54] "The Training Centre" (http://www.arsenal.com/the-club/training-centre). Arsenal F.C. . Retrieved 11 August 2008.

[55] "Get to Underhill Stadium" (http://www.arsenal.com/news/news-archive/how-to-get-to...-underhill-stadium). Arsenal F.C. . Retrieved 7 September 2008.

[56] Kempster, Tony. "Attendances 2007/08" (http://www.tonykempster.co.uk/prematt.htm?comp=1). . Retrieved 11 August 2008.

[57] "All Time League Attendance Records" (http://web.archive.org/web/20071029153705/http://www.nufc.com/html/attendance-all-time.html). NUFC.com. Archived from the original (http://www.nufc.com/html/attendance-all-time.html) on 29 October 2007. . Retrieved 11 August 2008. Please note that some pre-war attendance figures used by this source were estimates and may not be entirely accurate.

[58] "German Bundesliga Stats: Team Attendance – 2010–11" (http://soccernet.espn.go.com/stats/attendance/_/league/ger.1/year/2010/german-bundesliga?cc=5901). *ESPNsoccernet*. .

[59] http://arxiu.fcbarcelona.cat/web/english/noticies/club/temporada10-11/05/17/n110517117527.html

[60] "Barclays Premier League Stats: Team Attendance – 2010–11" (http://soccernet.espn.go.com/stats/attendance/_/league/eng.1/year/2010/barclays-premier-league?cc=5901). *ESPNsoccernet*. .

[61] "Spanish La Liga Stats: Team Attendance – 2010–11" (http://soccernet.espn.go.com/stats/attendance/_/league/ESP.1/year/2010/spanish-primera-division?cc=5901). *ESPNsoccernet*. .

[62] Noble, Kate (22 September 2002). "Boring, Boring Arsenal" (http://www.time.com/time/magazine/article/0,9171,353528,00.html). *Time*. . Retrieved 11 August 2008.

[63] "Fans Report 2006/2007" (http://www.arsenal.com/assets/_files/documents/jul_08/gun__1215526755_charter_report_20062007.doc) (Word document). Arsenal F.C. . Retrieved 7 September 2008.

[64] "Arsenal FC – the Premiership's fastest growing football brand" (http://web.archive.org/web/20070604205337/http://www.granadaventures.co.uk/newsdocs/doc28.doc) (Word document). Granada Ventures. 5 August 2005. Archived from the original (http://www.granadaventures.co.uk/newsdocs/doc28.doc) on 4 June 2007. .

[65] Coggin, Stewart. "The North London derby" (http://www.premierleague.com/page/Magazinedettail/0,,12306~1195275,00.html). Premier League. . Retrieved 7 September 2008.

[66] "The Classic: Arsenal-Manchester Utd" (http://www.fifa.com/worldfootball/clubfootball/news/newsid=109924.html). FIFA. 17 January 2007. . Retrieved 23 October 2009.

[67] "Club Rivalries Uncovered" (http://www.footballfanscensus.com/issueresults/Club_Rivalries_Uncovered_Results.pdf) (PDF). Football Fans Census. . Retrieved 7 September 2008.

[68] "Football Rivalries Report 2008" (http://www.footballpools.com/football-fever/rivalries-league.html). The New Football Pools.. Retrieved 7 September 2008.
[69] "Statement of Accounts and Annual Report 2006/2007 (p43)" (http://www.arsenal.com/assets/_files/documents/jul_08/gun__1215525940_Arsenal_Annual_Report_May_2007.pdf) (PDF). Arsenal Holdings plc. May 2007.. Retrieved 11 August 2008.
[70] "Statement of Accounts and Annual Report 2009/2010 (p41)" (http://www.arsenal.com/assets/_files/documents/sep_10/gun__1285660177_Arsenal_Holdings_plc_Annual_Re.pdf) (PDF). Arsenal Holdings plc. Sep 2010.. Retrieved 12 January 2011.
[71] http://www.plusmarketsgroup.com/details.shtml?ISIN=GB0030895238/GBX/PLUS-exn
[72] "Arsenal Holdings plc" (http://www.plusmarketsgroup.com/details.shtml?ISIN=GB0030895238/GBP/PLUS-exn). PLUS Markets Group.. Retrieved 5 November 2009.
[73] "Arsenal Holdings plc: 2009 Financial results" (http://www.arsenal.com/news/news-archive/arsenal-holdings-plc-results-for-year-end-ma). Arsenal Holdings plc. 28 September 2009.. Retrieved 5 November 2009.
[74] Scott, Matt and Allen, Katie (6 April 2007). "Takeover gains pace at Arsenal with 9.9% sale" (http://football.guardian.co.uk/News_Story/0,,2051488,00.html). *The Guardian*.. Retrieved 11 August 2008.
[75] "Kroenke increases stake in Arsenal Holdings" (http://www.arsenal.com/news/news-archive/kroenke-increases-stake-in-arsenal-holdings). Arsenal F.C. 5 November 2009.. Retrieved 6 December 2009.
[76] "Russian buys Dein's Arsenal stake" (http://news.bbc.co.uk/1/hi/business/6971124.stm). BBC News. 30 August 2007.. Retrieved 11 August 2008.
[77] "TR-1: Notifications of Major Interests in Shares" (http://www.plusmarketsgroup.com/story.shtml?ISIN=GB0030895238/GBX/PLUS-exn&NewsID=1234794458000759). PLUS Markets. 16 February 2009.. Retrieved 17 February 2009.
[78] "Kroenke joins Arsenal's Board of Directors" (http://www.arsenal.com/news/news-archive/other). Arsenal F.C. 19 September 2008.. Retrieved 19 September 2008.
[79] "Arsenal board announce revised 'lock-down' agreement" (http://www.arsenal.com/news/news-archive/board-announce-revised-lock-down-agreement). Arsenal F.C. 18 October 2007.. Retrieved 11 August 2008.
[80] http://sportshaze.com/england/english-premier-league/american-tycoon-in-advanced-talks-with-gunners-2826
[81] http://www.arsenal.com/the-club/corporate-info/the-arsenal-board
[82] *Firsts, Lasts & Onlys: Football* – Paul Donnelley (Hamlyn, 2010)
[83] "It Happened at Highbury: First live radio broadcast" (http://www.arsenal.com/news/news-archive/programme-it-happened-at-highbury). Arsenal F.C.. Retrieved 11 August 2008.
[84] "Happened on this day – 16 September" (http://news.bbc.co.uk/sport1/hi/funny_old_game/2260280.stm). BBC Sport. 16 September 2002.. Retrieved 11 August 2008.
[85] "History of Match of the Day" (http://news.bbc.co.uk/sport1/hi/tv_and_radio/match_of_the_day/879960.stm). BBC Sport. 14 February 2003.. Retrieved 11 August 2008.
[86] "Fans trial first live 3D sports event" (http://news.smh.com.au/breaking-news-sport/fans-trial-first-live-3d-sports-event-20100201-n7bo.html). *Sydney Morning Herald*. Associated Press. 1 February 2010.. Retrieved 16 March 2010.
[87] "The Arsenal Stadium Mystery" (http://www.imdb.com/title/tt0031055/). IMDb.. Retrieved 11 August 2008.
[88] "Arsenal at the movies" (http://www.arseweb.com/other/movies.html). Arseweb.. Retrieved 11 August 2008.
[89] "Nick Hornby" (http://www.guardian.co.uk/books/2008/jun/13/nick.hornby). *The Guardian*. 22 July 2008.. Retrieved 7 September 2008. "Critically acclaimed and commercial dynamite, Fever Pitch helped to make football trendy and explain its appeal to the soccerless"
[90] "Fever Pitch (1997)" (http://www.imdb.com/title/tt0119114/). IMDb.. Retrieved 7 September 2008.
[91] "Fever Pitch (2005)" (http://www.imdb.com/title/tt0332047/). IMDb.. Retrieved 7 September 2008.
[92] May, John (19 May 2003). "No more boring, boring Arsenal" (http://news.bbc.co.uk/sport1/hi/football/fa_cup/3037789.stm). BBC Sport.. Retrieved 7 September 2008.
[93] Association Football – Harry Enfield – Mr Cholmondley-Warner (http://www.youtube.com/watch?v=y4CXY6TVBMc&feature=related) YouTube
[94] "Laura Harvey" (http://www.arsenal.com/ladies/coaching-staff/laura-harvey). Arsenal FC.. Retrieved 7 July 2010.
[95] "Arsenal Ladies Honours" (http://www.arsenal.com/ladies/ladies-honours). Arsenal F.C.. Retrieved 21 May 2007.
[96] Mawhinney, Stuart (7 May 2007). "Arsenal clinch quadruple" (http://www.thefa.com/TheFACup/FACompetitions/TheFAWomensCup/NewsandFeatures/2009/FAWCMatchReport.aspx). The Football Association.. Retrieved 23 October 2009.
[97] "Get to Boreham Wood" (http://www.arsenal.com/ladies/boreham-wood-directions). Arsenal F.C.. Retrieved 7 September 2008.
[98] "Arsenal Charity Ball raises over £60,000" (http://www.arsenal.com/news/news-archive/arsenal-charity-ball-raises-over-pound-60-000). Arsenal F.C. 11 May 2006.. Retrieved 11 August 2008.
[99] "Ex-Pro and Celebrity XI" (http://web.archive.org/web/20080822031124/http://www.arsenal.com/the-club/community/ex-pro-and-celebrity-xi). Arsenal F.C. Archived from the original (http://www.arsenal.com/the-club/community/ex-pro-and-celebrity-xi) on 22 August 2008.. Retrieved 11 August 2008.
[100] Arsenal smash fundraising target for GOSH (http://www.arsenal.com/news/community-news/arsenal-smash-fundraising-target-for-gosh) Arsenal FC, 2 August 2010
[101] "Arsenal force Spanish hat shop to change name" (http://www.insideworldsoccer.com/2011/09/arsenal-force-spanish-hat-shop-to.html). inside World Soccer. 9 September 2011..
[102] "Club Records" (http://www.arsenal.com/history/club-records). Arsenal F.C.. Retrieved 23 October 2009.

[103] "First Team Players: Thierry Henry" (http://www.arsenal.com/first-team/players/thierry-henry). Arsenal.com. . Retrieved 11 February 2012.
[104] "Wright salutes Henry's goal feat" (http://news.bbc.co.uk/sport1/hi/football/teams/a/arsenal/4356664.stm). BBC Sport. 19 October 2005. . Retrieved 11 August 2008.
[105] Ward, Rupert. "Arsenal vs Bolton. 13/09/97" (http://www.arseweb.com/97-98/reports/130997.html). Arseweb. . Retrieved 11 August 2008.
[106] "Arsenal 2–3 West Ham" (http://news.bbc.co.uk/sport1/hi/football/eng_prem/4657206.stm). BBC Sport. 1 February 2006. . Retrieved 23 October 2009.
[107] "Man Utd game attracts record attendance" (http://www.arsenal.com/news/news-archive/man-utd-game-attracts-record-attendance). Arsenal F.C. 5 November 2007. . Retrieved 11 August 2008.
[108] "First Team Players" (http://www.arsenal.com/first-team/players). Arsenal F.C.. . Retrieved 6 January 2012.
[109] "Reserve Players" (http://www.arsenal.com/reserves-youth/players). Arsenal F.C.. . Retrieved 13 August 2009.
[110] "Arsenal FC" (http://www.uefa.com/uefachampionsleague/season=2012/clubs/club=52280/index.html). UEFA. . Retrieved 9 August 2011.
[111] "Andrey Arshavin rejoins Zenit St Petersburg from Arsenal on loan" (http://www.bbc.co.uk/sport/0/football/17160447). BBC Football. 24 February 2012. . Retrieved 24 February 2012.
[112] "Vito Signs In With City" (http://www.hullcityafc.net/page/NewsDetail/0,,10338~2566231,00.html). Hull City A.F.C. 4 January 2012. . Retrieved 4 January 2012.
[113] "Ryo joins Bolton Wanderers on loan" (http://www.arsenal.com/news/news-archive/ryo-joins-bolton-wanderers-on-loan). Arsenal F.C.. 31 January 2012. . Retrieved 31 January 2012.
[114] "Afobe joins Reading on loan for the season" (http://www.arsenal.com/news/news-archive/afobe-joins-reading-on-loan-for-the-season). Arsenal F.C.. 22 March 2012. . Retrieved 22 March 2012.
[115] "Aneke joins Preston until end of the season" (http://www.arsenal.com/news/reserves-news/aneke-joins-preston-until-end-of-the-season). Arsenal F.C.. 22 March 2012. . Retrieved 22 March 2012.
[116] "Wycombe Wanderers sign Craig Eastmond and Chris Hackett on loan" (http://www.bbc.co.uk/sport/0/football/17115081). BBC Sport. 21 February 2012. .
[117] "Hajrovic extends loan spell with Barnet" (http://www.arsenal.com/news/news-archive/hajrovic-extends-loan-spell-with-barnet). Arsenal F.C.. 4 April 2012. . Retrieved 7 April 2012.
[118] "Lansbury moves on loan to West Ham United" (http://www.arsenal.com/news/news-archive/lansbury-moves-on-loan-to-west-ham-united). 31 August 2011. . Retrieved 31 August 2011.
[119] "Murphy joins Preston North End on loan" (http://www.arsenal.com/news/reserves-news/murphy-joins-preston-north-end-on-loan). 30 January 2012. . Retrieved 30 January 2012.
[120] "Nicklas Bendtner loaned to Sunderland" (http://www.arsenal.com/news/news-archive/nicklas-bendtner-loaned-to-sunderland). 31 August 2011. . Retrieved 31 August 2011.
[121] "Sanchez Watt joins Crawley Town on loan" (http://www.arsenal.com/news/reserves-news/sanchez-watt-joins-crawley-town-on-loan). 26 January 2012. . Retrieved 29 January 2012.
[122] "Yennaris joins Notts County on loan" (http://www.arsenal.com/news/news-archive/yennaris-joins-notts-county-on-loan). 23 March 2012. . Retrieved 23 March 2012.
[123] "Ally Welcomes Kyle Return" (http://www.rangers.co.uk/articles/20110804/ally-welcomes-kyle-return_2254024_2410614). Rangers.co.uk. 4 August 2011. . Retrieved 4 August 2011.
[124] "Campbell and Sunu heading for Lorient" (http://www.arsenal.com/news/news-archive/campbell-and-sunu-heading-for-lorient). 31 August 2011. .
[125] "Denilson returns to Brazil for Sao Paulo loan" (http://www.arsenal.com/news/news-archive/denilson-returns-to-brazil-for-sao-paulo-loan). Arsenal.com. 20 July 2011. . Retrieved 20 July 2011.
[126] "Vela and Botelho go on loan to Spain" (http://www.arsenal.com/news/news-archive/vela-and-botelho-go-on-loan-to-spain). 16 August 2011. . Retrieved 16 August 2011.
[127] "First Team Coaching Staff" (http://www.arsenal.com/first-team/coaching-staff#Coaching Staff). Arsenal F.C. . Retrieved 23 October 2009.
[128] "Reserves & Youth Coaching Staff" (http://www.arsenal.com/reserves-youth/coaching-staff#Coaching Staff). Arsenal F.C. . Retrieved 23 October 2009.
[129] Ducker, James (5 September 2009). "Scouting networks extend search for talent all over the world" (http://www.timesonline.co.uk/tol/sport/football/article6822692.ece). *The Times* (UK). . Retrieved 23 October 2009.
[130] Soar & Tyler (2005). *The Official Illustrated History of Arsenal*. p. 30.
[131] "Wenger hungry for further success" (http://news.bbc.co.uk/sport1/hi/football/eng_prem/8286929.stm). BBC Sport. 2 October 2009. . Retrieved 6 December 2009.
[132] Davies, Christopher (8 December 2008). "700 and not out: Arsenal boss Wenger looking to celebrate memorable day at Porto" (http://www.dailymail.co.uk/sport/football/article-1092959/700-Arsenal-boss-Wenger-looking-celebrate-memorable-day-Porto.html). *Daily Mail* (UK). . Retrieved 6 December 2009.
[133] "The Managers" (http://www.arsenal.com/history/the-managers). Arsenal F.C. . Retrieved 7 September 2008.

[134] Up until 1992, the top division of English football was the Football League First Division; since then, it has been the Premier League. Similarly until 1992, the Second Division was the second tier of league football, which since the 2004–05 season has been known as The Championship.

[135] The trophy was known as the Charity Shield until 2001, and as the Community Shield since then. Reference: Ross, James (28 August 2009). "England – List of FA Charity/Community Shield Matches" (http://www.rsssf.com/tablese/engsupcuphist.html). RSSSF. . Retrieved 12 January 2010.

[136] Ross, James M (28 August 2009). "England – List of Champions" (http://www.rsssf.com/tablese/engchamp.html#sall). RSSSF. . Retrieved 23 October 2009.

[137] Ross, James M (12 June 2009). "England FA Challenge Cup Finals" (http://www.rsssf.com/tablese/engcuphist.html#sall). RSSSF. . Retrieved 23 October 2009.

[138] Stokkermans, Karel (24 September 2009). "Doing the Double: Countrywise Records" (http://www.rsssf.com/miscellaneous/doublerec.html#coun). RSSSF. . Retrieved 23 October 2009.

[139] Collins, Roy (20 May 2007). "Mourinho collects his consolation prize" (http://www.telegraph.co.uk/sport/football/2313666/Mourinho-collects-his-consolation-prize.html). *The Daily Telegraph*. . Retrieved 6 December 2009. "Chelsea's Cup came wrapped in an extra ribbon, only the second team after Arsenal in 1993 to win both domestic cups."

[140] "Arsenal Football Club" (http://www.premierleague.com/page/arsenal). Premier League. . Retrieved 11 August 2008.

[141] Ross, James M (12 June 2009). "England FA Challenge Cup Finals" (http://www.rsssf.com/tablese/engcuphist.html). RSSSF. . Retrieved 23 October 2009.

[142] "The Unbeaten Record" (http://www.arsenal.com/history/club-records/the-unbeaten-record). Arsenal Broadband. . Retrieved 10 December 2011.

Further reading

- Hayes, Dean (2007). *Arsenal: The Football Facts*. John Blake. ISBN 978-1-84454-433-2.
- Hornby, Nick (1992). *Fever Pitch*. Indigo. ISBN 978-0-575-40015-3.
- Maidment, Jem (2006). *The Official Arsenal Encyclopedia*. Hamlyn. ISBN 978-0-600-61549-1.
- Soar, Phil & Tyler, Martin (2000). *The Official Illustrated History of Arsenal*. Hamlyn. ISBN 978-0-600-60175-3.
- Spurling, Jon (2004). *Rebels for the Cause: The Alternative History of Arsenal Football Club*. Mainstream. ISBN 978-1-84018-900-1.
- Stammers, Steve (2008). *Arsenal: The Official Biography – The Compelling Story of an Amazing Club*. Hamlyn. ISBN 978-0-60061-892-8.

External links

Official websites

- Official website (http://http://www.arsenal.com/) (Chinese) (English) (Japanese) (Korean)
- Arsenal (http://www.premierleague.com/page/arsenal/0,,12306~1072252,00.html) at the Premier League official website
- Arsenal (http://www.uefa.com/footballEurope/Club=52280/competition=1/index.html) at the UEFA official website

News sites

- Arsenal on BBC Sport: Club news (http://news.bbc.co.uk/sport1/hi/football/teams/a/arsenal/default.stm) – Recent results (http://news.bbc.co.uk/sport1/hi/football/teams/a/arsenal/results/default.stm) – Upcoming fixtures (http://news.bbc.co.uk/sport1/hi/football/teams/a/arsenal/fixtures/default.stm) – Club statistics (http://news.bbc.co.uk/sport1/hi/football/teams/a/arsenal/club_stats/default.stm)
- Arsenal news (http://www.skysports.com/football/teams/arsenal) from Sky Sports
- Arsenal News (http://www.arsenalnews.co.uk/)

Other

- Arsenal F.C. companies (http://opencorporates.com/corporate_groupings/Arsenal+Football+Club) grouped at OpenCorporates

Club_Brugge_K.V.

Full name	Club Brugge Koninklijke Voetbalvereniging (*Club Bruges Royal Football association*)
Nickname(s)	*Blauw-Zwart* (Blue-Black), *Club, FCB*
Founded	November 13, 1891 Stamnummer (matricule number) 3
Ground	Jan Breydelstadion (Jan Breydel Stadium), Bruges (Capacity: 29,472[1])
President	Bart Verhaeghe
Head coach	Christoph Daum
League	Belgian Pro League
2010–11	Belgian Pro League, 4th

Home colours Away colours

Club Brugge Koninklijke Voetbalvereniging (Dutch pronunciation: [klʏˈbrʏɟə ˈkoː.nɪŋk.lə.kə ˈvut.bɑɫ.və.ˌreːnə.ɟɪŋ]) is a football club from Bruges in Belgium. It was founded in 1891 and is one of the top clubs in Belgium. Its home ground is the Jan Breydel Stadium, which has a capacity of 29,472.

Jan Breydel Stadium

Club Brugge's major rival is R.S.C. Anderlecht, and it shares the Jan Breydel Stadium with city rival Cercle Brugge K.S.V., with whom they contest the Bruges derby.

Throughout its long history, Club Brugge has enjoyed much European football success, reaching two European finals and two European semi-finals. Club Brugge is the only Belgian club to have played the final of the European Cup (forerunner of the current UEFA Champions League) so far. They were beaten by Liverpool F.C. in the final of its 1978 season. They also lost in the 1976 UEFA Cup Final to Liverpool. Club Brugge holds the record number of consecutive participations in the UEFA Cup (16).

Tifo before the Champions League game Club Brugge-Rapid Wien in 2005

History

- **1890: Brugsche Football Club**

Club created by old students of the Catholic school Broeders Xaverianen and the neutral school Koninklijk Atheneum.

- **13 November 1891: Club recreated**

The club was recreated. This has since been adopted as the official date of foundation.

- **1892: First board**

An official board was installed in the club.

- **1894: Football Club Brugeois**

Club created by 16 old members of Brugsche FC.

- **1895: Vlaamsche Football Club de Bruges**

Club created in the city.

- **1895/1896: the UBSSA set up in 1895. and they went to the UBSSA and took part of the first Belgian national league.**
- **1896: Leaving the UBSSA**

Financial it was difficult for FC Brugeois and so after only one year they had to leave the UBSSA.

- **1897: Fusion**

FC Brugeois joined Brugsche FC but they continued under the name Football Club Brugeois.

- **1902: New fusion**

Vlaamsche FC joined FC Brugeois.

- **1912: De Klokke**

They moved to a new stadium named "De Klokke".

- **1913/1914: First cup final**

FC Brugeois reached their first Belgian Cup final but they lost 2–1 from Union SG.

- **1920: First time league champions**

The club became for the first time champions of the first division.

- **1926: Royal Football Club Brugeois**

The club get number 3 as their matricule number and in the same year they get the royal title.

- **1928: First relegation**

A first low when the club was relegated to the second division.

- **1930: New statute**

President Albert Dyserynck changed the club's statute into a non-profit association.

- **1931: Albert Dyserynckstadion**

When president Albert Dyserynck suddenly died they honoured him by changing the stadium's name into Albert Dyserynckstadion.

- **1959: Permanent to the first division**

RFC Brugeois promoted to the first division and never relegated again in the future.

- **1968: First time cup winners**

They won the Belgian Cup for the first time against Beerschot AC (1–1, 7–6 after penalty's).

- **1972: Club Brugge Koninklijke Voetbalvereniging**

The club changed their name into the Flemisch name Club Brugge KV

- **1975: Olympiastadion**

They moved from Albert Dyserynckstadion to Olympiastadion (current Jan Breydelstadion).

- **1976: Highest position in UEFA Cup**

Under Austrian coach Ernst Happel, Club Brugge reached the finals of the UEFA Cup and lost against Liverpool (3–2 and 1–1).

- **1978: Only Belgian European Cup 1 finalists**

Still under Ernst Happel, the club faced Liverpool again of a European final. This time it was in the European Champions Clubs' Cup final. And again they lost (1–0). Club Brugge is the only Belgian club that have reached the finals of the European biggest competition.

- **1992: First goal scorer in the Champions League**

Daniel Amokachi is the first goal scorer in the Champions League. He scored against CSKA Moskva.

- **1998: Jan Breydelstadion**

Olympiastadion had to be expand for the EURO 2000 organisation. They also changed the name into Jan Breydelstadion.

- **2006: CLUBtv**

It was the first Belgian club to create its own TV channel.

Honours

Domestic

- **Belgian First Division champions: 13**

  1919–20, 1972–73, 1975–76, 1976–77, 1977–78, 1979–80, 1987–88, 1989–90, 1991–92, 1995–96, 1997–98, 2002–03, 2004–05

- **Belgian Cup winners: 10**

  1967–68, 1969–70, 1976–77, 1985–86, 1990–91, 1994–95, 1995–96, 2001–02, 2003–04, 2006–07

- **Belgian Supercup winners: 13**

  1980, 1986, 1988, 1990, 1991, 1992, 1994, 1996, 1998, 2002, 2003, 2004, 2005

International

- **European Champion Clubs' Cup quarter finalist: 1**

  1976–77

- **European Champion Clubs' Cup finalist: 1**

  1977–78

- **UEFA Cup finalist: 1**

  1975–76

- **UEFA Cup semi finalist: 1**

  1987–88

- **Bruges Matins winners: 17**

  1979, 1981, 1984, 1990, 1992, 1993, 1995, 1996, 1998, 2000, 2001, 2004, 2006, 2007, 2008, 2009, 2011

- **Kirin Cup winners: 1**

  1981

European record

As of August 19, 2011.

Competition	A	GP	W	D	L	GF	GA
European Cup / UEFA Champions League	14	77	32	16	29	106	93
UEFA Cup Winners' Cup	6	28	15	3	10	41	33
UEFA Cup / UEFA Europa League	24	126	59	26	41	226	166

A = appearances, GP = games played, W = won, D = drawn, L = lost, GF = goals for, GA = goals against.

Matches

- Q = qualification round
- R = round
- Group = group stage / Group 1 = first group stage / Group 2 = second group stage
- 1/8 = eighth finals / 1/4 = quarter-finals / 1/2 = semi-finals
- F = final
- PUC = points UEFA coefficient

Total points for UEFA coefficient: 261.0.

Summary of best results

From the quarter-finals upwards:

(2 finals)

European Cup/UEFA Champions League *(1)*:

- *finalists in 1978*
- quarter-finalists in 1977

UEFA Cup Winners' Cup:

- semi-finalists in 1992
- quarter-finalists in 1971 and 1995

UEFA Cup/UEFA Europa League *(1)*:

- *finalists in 1976*
- semi-finalists in 1988

UEFA club coefficient ranking

(As of 12 November 2011), Source: uefa.com website [2]

Rank	Team	Points
56	LOSC Lille	34.968
57	**Club Brugge**	31.880
58	AZ Alkmaar	31.863

Players

First team squad

As of February 1, 2012. Note: Flags indicate national team as has been defined under FIFA eligibility rules. Players may hold more than one non-FIFA nationality.

No.	Position	Player
1	GK	Colin Coosemans
2	DF	Tom Høgli
3	DF	Michael Almebäck
4	DF	Carl Hoefkens *(captain)*
5	DF	Fredrik Stenman
6	MF	Niki Zimling
7	FW	Carlos Bacca
8	MF	Lior Refaelov
9	FW	Björn Vleminckx
11	MF	Jonathan Blondel
13	MF	Víctor Vázquez
15	FW	Joseph Akpala
16	MF	Maxime Lestienne

No.	Position	Player
17	DF	Marcos Camozzato
18	DF	Ryan Donk *(vice-captain)*

No.	Position	Player
19	MF	Thomas Meunier
20	MF	Thibaut Van Acker
21	DF	Pietro Perdichizzi
22	DF	Jordi
24	DF	Daan van Gijseghem
25	FW	Mushaga Bakenga
26	MF	Fries Deschilder
27	MF	Jimmy De Jonghe
28	DF	Jannes Vansteenkiste
29	FW	Zinho Gano
30	FW	Nick Van Belle
32	MF	Vadis Odjidja-Ofoe
33	GK	Vladan Kujović
39	GK	Bojan Jorgačević
50	GK	Sven Dhoest

For recent transfers, see the list of Belgian football transfers summer 2011.

Out on loan

Note: Flags indicate national team as has been defined under FIFA eligibility rules. Players may hold more than one non-FIFA nationality.

No.	Position	Player
25	DF	Júnior Díaz *(at Wisła Kraków until 30 June 2012)*
40	FW	Dorge Kouemaha *(at 1. FC Kaiserslautern until 30 June 2012)*

Reserve team squad

As of June 6, 2011. Note: Flags indicate national team as has been defined under FIFA eligibility rules. Players may hold more than one non-FIFA nationality.

No.	Position	Player
22	MF	Birger Verstraete
45	FW	Tjendo De Cuyper
49	GK	Jordy Maes
51	DF	Mateusz Kuzmicki [1]
52	DF	Simon Savaete
53	DF	Dylan Vanneste
54	MF	Nathaniel Lagrou
55	FW	Marijn Vandewalle
60	DF	Björn Engels
	GK	Ilias Rchidi
	DF	Robin De Feyter

No.	Position	Player
	DF	Maxime Gunst
	DF	Mathias Moeyersons
	MF	Rudi Chirishungu
	MF	Gaetan Lambrechts
	MF	Bram Leroy
	MF	Mick Michiels
	MF	Jaric Schaessens
	MF	Matthias Schaessens
	MF	Jaan Vanwildemeersch
	FW	Jordan Botaka [2]

Notes:

1. Has dual citizenship; second is
2. Has dual citizenship; second is

Club captains

- Jan Ceulemans (1986–91)
- Franky Van der Elst (1991–99)
- Vital Borkelmans (1999-00)
- Gert Verheyen (2000–02)
- Dany Verlinden (2002–04)
- Timmy Simons (2004–05)
- Gert Verheyen (2005–06)
- Sven Vermant (2006–08)
- Philippe Clement (2008–09)
- Stijn Stijnen (2009–10)
- Carl Hoefkens (2010–)

Retired numbers

12 – The 12th man (reserved for the club supporters)

23 – In recognition of striker François Sterchele. Sterchele died in a single-person car accident on May 8, 2008.

Club officials

Honorary presidents and directors

- Dr. Michel D'Hooghe (Honorary president)
- Fernand De Clerck (Honorary president)
- Marcel Kyndt (Honorary vice-president)
- Raoul Beuls (Honorary vice-president)
- Chris Caestecker (Honorary director)
- Guido Claeys (Honorary director)
- Dr. William De Groote (Honorary director)
- Guy Jacobs (Honorary director)
- Pol Jonckheere (Honorary director)
- André Piccu (Honorary director)
- Herman Valcke (Honorary director)
- Hugo Vandamme (Honorary director)
- Dr. Roland Watteyne (Honorary director)

Board

- Bart Verhaeghe (President)
- Vincent Mannaert (Managing director)
- Jan Boone (Director)
- Bart Coeman (Director)
- Sam Sabbe (Director)
- Ignace Van Doorselaere (Director)

Presidential history

- Philippe Delescluze (1891-00)
- Albert Seligmann (1900–02)
- Alfons De Meulemeester (1903–14)
- Albert Dyserynck (1919–31)
- Fernand Hanssens (1932–37)
- Emile De Clerck (1937–59)
- André De Clerck (1959–73)
- Fernand De Clerck (1973–99)
- Michel Van Maele (1999-03)
- Dr. Michel D'Hooghe (2003–09)
- Pol Jonckheere (2009–11)
- Bart Verhaeghe (2011–)

Management

- Vincent Mannaert (General manager)
- Veroniek Degrande (Finance manager)
- Wim De Meyer (Communication & PR manager)
- Jacques De Nolf (General secretary)
- Kristof Lemahieu (General counsel)
- Henk Mariman (Sports manager)
- Patrick Orlans (Commercial manager)
- Sven Vermant (Sports manager)

Coaching staff

- Christoph Daum (Head coach)
- Rudi Verkempinck (Assistant coach)
- Dany Verlinden (Coach goalkeepers)
- Joost Desender (Physical coach)
- Siebe Hannoset (Performance coach)
- Philippe Clement (Line coach defenders)
- Stéphane Van Der Heyden (Line coach midfielders)
- Kenneth Brylle Larsen (Line coach strikers)

Head coaches history

- Hector Goetinck (1930–33)
- Gérard Delbeke (1933–34)
- Arthur Volckaert (1934–36)
- Karl Schrenk (1936–38)
- Robert De Veen (1938–39)
- Gérard Delbeke (1939–45)
- Louis Versyp (1945–50)
- William Kennedy (1950–51)
- Félix Schavy (1951–57)
- Norberto Höfling (1957–63)
- Juan Schwanner (1963)
- Henri Dekens (1963–65)
- Ludwig Dupal (1965–67)
- Norberto Höfling (1967–68)
- Milorad Pavić (1968–69)
- Frans de Munck (1969–71)
- Leo Canjels (1971–73)
- Jaak De Wit (1973–74)
- Ernst Happel (1974–78)
- András Béres (1978–79)
- Mathieu Bollen (1979)
- Han Grijzenhout (1979–80)
- Gilbert Gress (1980–81)
- Antoine "Spitz" Kohn (1981)
- Henri Coppens (1981–82)
- Raymond Mertens (1981–82)
- Georg Keßler (1982–84)
- Henk Houwaart (1984–89)
- Georges Leekens (1989–91)
- Hugo Broos (1991–97)
- Eric Gerets (1997–99)
- René Verheyen (1999–00)
- Trond Sollied (2000–05)
- Jan Ceulemans (2005–06)
- Emilio Ferrera (2006–07)
- Čedomir Janevski (2007)
- Jacky Mathijssen (2007–09)
- Adrie Koster (2009–11)
- Christoph Daum (2011–..)

Medical staff

- Dr. Roland Watteyne (Honorary director)
- Dr. Jan De Neve (Head of medical services)
- Dr. Kris Vandecasteele (Club doctor)
- Dr. Pieter D'Hooghe (Club doctor/Orthopedic consultant)
- Geert Ryckebusch (Physiotherapist)
- Dimitri Dobbenie (Physiotherapist)
- Hans Kerckaert (Physiotherapist)
- Thomas Geschier (Physiotherapist)

Logistical staff

- Lode Lobbestael (Team official)
- Roger Tierenteyn (Logistical staff member)
- Pascal Plovie (Logistical staff member)

Reserves coaching staff

- Philippe Clement (Head coach)
- Jannes Tant (Assistant coach)
- Dirk Laleman (Physical coach)
- Hans Kerckaert (Physiotherapist)
- Kristoff Deryckere (Team mentor)
- Michel Dierings (Team mentor)

References

[1] http://www.uefa.com/MultimediaFiles/Download/StatDoc/competitions/-Publications/01/67/58/96/1675896_DOWNLOAD.pdf
[2] http://www.uefa.com/memberassociations/uefarankings/club/index.html

External links

- Official Website (http://www.clubbrugge.be) (**Dutch**) / (**French**) / (**English**) / (**Spanish**)
- Official Facebook Fanpage (http://www.facebook.com/clubbrugge)
- Official Fan Federation (http://www.supportersfederatieclubbrugge.be) (**Dutch**)
- Blue Army – Fan Association & Fan-zine (http://www.blue-army.com) (**Dutch**)
- Club Brügge KV XtraTime Fanpage (http://www.xtratime.org/forum/showthread.php?t=216435) (**English**)
- Club Brügge at UEFA.COM (http://www.uefa.com/footballEurope/Club=50043/domestic.html) (**English**)
- Club Brügge at EUFO.DE (http://www.eufo.de/football/bel/clubugge.htm)
- Club Brügge at Weltfussball.de (http://www.weltfussball.de/teams/club-brugge-kv/)
- Club Brügge at Playerhistory.com (http://www.playerhistory.com/club/20/)
- Club Brügge at Transfermarkt.de (http://www.transfermarkt.de/de/verein/2282/fc-bruegge/uebersicht/startseite.html)
- Club Brügge at Football Squads.co.uk (http://www.footballsquads.co.uk/belgium/2010-2011/prolge/clubbrug.htm)
- Club Brügge at National Football Teams.com (http://www.national-football-teams.com/v2/club.php?id=222)
- Club Brügge at Football Lineups.com (http://www.football-lineups.com/team/Club_Brugge/)

K.V._Kortrijk

Full name	Koninklijke Voetbalclub Kortrijk
Nickname(s)	De Kerels
Founded	1901
Ground	Guldensporenstadion, Kortrijk (Capacity: 9,399)
Chairman	Joseph Allijns
Manager	Hein Vanhaezebrouck
League	Belgian Pro League
2010-11	Belgian Pro League, 10th Group B Europa League Playoff, 4th

Home colours **Away colours**

Koninklijke Voetbalclub Kortrijk (often simply called **Kortrijk** or **KVK**) is a Belgian professional football club based in Kortrijk, West Flanders. They have been playing in the Belgian Pro League since the 2008-09 season, and they achieved their best ranking ever during the 2009-10 season, finishing 4th after the play-offs. Kortrijk was founded in 1971, though their roots can be traced to 1901. They are registered to the Royal Belgian Football Association with matricule number 19. The club colours are red and white. They play their home matches at the Guldensporenstadion, named after the Battle of the Golden Spurs which took place in Kortrijk in 1302.

Kortrijk enjoyed a first spell in the first division in the early 20th century, between 1906-07 and 1910-11. They eventually had to wait 65 years to return at the highest level, when they enjoyed 15 seasons at the top flight, interrupted by a season in the second division (between 1976-77 and 1978-79 and between 1980-81 and 1991-92). Kortrijk also played the 1998-99 season in the first division before they returned in 2008-09.

History

In 1901 SC Courtraisien was founded. The club merged with FC Courtraisien in 1918. They received in 1951 the name **Koninklijke Kortrijk Sport**. They had the matricule n°19 like the oldest club SC Courtraisien. **Stade Kortrijk** was founded in 1923 with matricule n°161. Because Stade Kortrijk and Koninklijke Kortrijk Sport didn't play well anymore, they merged in 1971. They could play with matricule n°19 and the new name was **KV Kortrijk**.

Due to a bankruptcy, KV Kortrijk had to relegate in 2001 to the third division. From 2004 to 2008 KV Kortrijk played in the second division, however in the 2007-08 season they managed to get promoted to the first division. The club is also home to the majority of Swansea City A.F.C. fans from Belgium.

Honours

- **Belgian Cup:**
  - **Runners-up (1):** 2011–12
- **Belgian Second Division:**
  - **Winners (2):** 1905–06, 2007–08
  - **Runners-up (2):** 1979–80, 1997–98
- **Belgian Second Division Final Round:**
  - **Winners (3):** 1976, 1980, 1998

Colours & badge

KV Kortrijk's colours are generally red and white. The home kit is usually all red with a white trim, and red shorts. The away kit is similar, except in blue. The club's logo is a white shield, split into thirds by a red upside-down, V-like shape, and the letters "*KVK*" are split into each third, in a red font. On top of the shield is a crown.

Stadium

KV Kortrijk play their home matches at the Guldensporen Stadion. It is located in Kortrijk, Belgium, and the current capacity of the ground is 9,399. In the summer of 2008, the grounds capacity was increased from 6,896 to 9,399, after Kortrijk's elevation into the Belgian First Division. *Guldensporenstadion* literally means "Stadium of the Golden Spurs". The stadium is named after this in reference to the Battle of the Golden Spurs which, in 1302, was fought in Kortrijk. The stadium has a traditional local stadium look about it, and is single-tiered all around the ground.

Current squad

As of 25 July, 2011

Note: Flags indicate national team as has been defined under FIFA eligibility rules. Players may hold more than one non-FIFA nationality.

No.	Position	Player
1	GK	Kristof van Hout
2	DF	*David Vandenbroek (on loan from Zulte-Waregem)*
3	DF	Baptiste Martin
4	MF	Ismaïla N'Diaye
5	DF	*Alassane També*
6	DF	*David Wijns*
7	FW	Steven Joseph-Monrose
8	MF	Nebojša Pavlović
9	FW	Dalibor Veselinović *(on loan from Anderlecht)*
10	FW	Ernest Nfor
11	DF	Mustapha Oussalah

No.	Position	Player
12	MF	*Geoffrey Ghesquire*
14	DF	Landry Mulemo
15	FW	*Dieter Meulenyzer*
16	GK	Darren Keet
17	MF	Gertjan De Mets
18	FW	Pablo Chavarría *(on loan from Anderlecht)*
19	MF	*Brecht Dejaegere*
20	MF	Mohamed Messoudi
21	MF	Brecht Capon
23	DF	Ervin Zukanović
77	MF	Péter Czvitkovics

Players out on loan

Note: Flags indicate national team as has been defined under FIFA eligibility rules. Players may hold more than one non-FIFA nationality.

No.	Position	Player
12	FW	Ebrahim Savaneh *(at RAEC Mons)*

Famous former players

- Émile Mpenza
- Mbo Mpenza
- Christian Benteke
- Xavier Chen
- Laurent Ciman
- Davy De Beule
- Michel De Wolf
- Philippe Desmet
- Jimmy Hempte
- Sven Kums
- Stijn Meert
- Axel Smeets
- Lorenzo Staelens
- Stefaan Tanghe
- David Vandenbroeck
- Glenn Verbauwhede
- Adlène Guedioura
- Djamel Zidane
- Aurelio Vidmar
- Petar Aleksandrov
- Aloys Nong
- Karim Belhocine
- Cédric Bétrémieux
- William Prunier
- Foeke Booy
- Anton Janssen
- Hendrie Krüzen
- Elimane Coulibaly
- Salif Keita
- Cheikhou Kouyaté
- Damir Desnica

Managers

- André Van Maldeghem & Luc Maddens (1971-73)
- André Van Maldeghem & Walter Maes (1973-75)
- Georges Heylens (1975-77)
- Raymond Mertens (1977-79)
- Henk Houwaart (1979-83)
- Wim Reijers (1983-84)
- André Van Maldeghem (1984-85)
- Dimitri Davidović (1985-86)
- Aad Koudijzer (1986)
- Han Grijzenhout (1986-87)
- Eddy Kinsabil (1987-88)
- Georges Leekens (1988-89)
- Henk Houwaart (1989-90)
- René Desayere (1990-91)
- Boudewijn Braem (1991-92)
- Johan Boskamp (1992-93)
- James Storme (1993-94)
- Patrick Van Geem (1994-95)
- Regi Van Acker (1995-98)
- Michel De Wolf (1998-99)
- Luc Vanderschommen (1999-00)
- Eddy Kinsabil (2000)
- Gerrit Laverge (2000-01)
- Boudewijn Braem (2001-02)
- Francky Dekenne (2002-03)
- Boudewijn Braem (2003)
- Angelo Nijskens (2003-04)
- Manu Ferrera (2004)
- Rudi Verkempinck (2004-05)
- Manu Ferrera (2005-06)
- Hein Vanhaezebrouck (2006-09)
- Georges Leekens (2009-10)
- Hein Vanhaezebrouck (2010-)

External links

- Official Website [1]
- Fansite [2]
- Kortrijk at UEFA.COM [3]
- Kortrijk at EUFO.DE [4]
- Kortrijk at Weltfussball.de [5]
- Kortrijk at Playerhistory.com [6]
- Kortrijk at Transfermarkt.de [7]
- Kortrijk at Football Squads.co.uk [8]
- Kortrijk at National Football Teams.com [9]
- Kortrijk at Football-Lineups.com [10]

References

[1] http://www.kvkortrijk.be/
[2] http://www.kvkortrijk.net/
[3] http://www.uefa.com/footballeurope/club=68495/domestic.html
[4] http://www.eufo.de/football/bel/kv_krijk.htm
[5] http://www.weltfussball.de/teams/kv-kortrijk/
[6] http://www.playerhistory.com/club/2213/
[7] http://www.transfermarkt.de/de/verein/601/kv-kortrijk/uebersicht/startseite.html
[8] http://www.footballsquads.co.uk/belgium/2008-2009/eerste/kortrijk.htm
[9] http://www.national-football-teams.com/v2/club.php?id=252
[10] http://www.football-lineups.com/team/Kortrijk/

Football_in_Belgium

Football, a sport which has been played in Belgium since the end of the 19th century, is that country's most popular sport. The national association was founded in 1895 with the intention of bringing some order and organization to the sport. The first match of the national team was played on May 1, 1904 (3-3 against France).

The teams with the highest average attendances during the 2009-2010 season were:

1 Standard de Liège - 24.406 - max. 30.000
2 Club Brugge KV - 24.368 - max. 28.151
3 RSC Anderlecht - 23.054 - max. 24.000
4 KRC Genk - 19.604 - max. 21.468
Traditionally, Anderlecht, Club Brugge and Standard de Liege are the three big teams, all of them playing and/or winning 1 or more European Cup Final(s).

National style

Both the national football team and the top Belgium division have a reputation for physical play. This came as a result of a lack of technically skilled foreign players allowed to play in Belgium due to legal restrictions. This changed after the Bosman ruling which forced the liberalization of the football player market in Europe. In response, Belgian clubs began to buy unknown players from Eastern Europe, South America and Africa. This had two contradictory consequences. On the one hand, the national team was weakened by the reduced opportunity for native Belgium players to gain a spot on domestic teams. On the other hand, the Jupiler League reinforced its status as an entry league for players who then move on to some of the greatest European clubs.
Indeed, some of the most talented players in Europe have played in Belgian clubs: Jean-Pierre Papin, Daniel Amokachi, Antolín Alcaraz and David Rozehnal were discovered at Club Brugge; Sunday Oliseh and Victor Ikpeba at RFC Liégeois; Jan Koller, Nii Lamptey and Aruna Dindane at R.S.C. Anderlecht; and Mido at K.A.A. Gent.
Others to have started or launched their careers in Belgium include Emmanuel Eboue, Romaric, Gervinho, Didier Zokora, Arthur Boka, Ivica Dragutinovic, Mario Stanic, Morten Olsen, Dorinel Munteanu, André Cruz, Seol Ki-Hyeon, Kennet Andersson, Klas Ingesson, Aaron Mokoena, Michaël Ciani, Nicolas Pareja, Oguchi Onyewu, Rabiu Afolabi, Cheick Tioté, Peter Odemwingie, Joseph Yobo, Ouwo Moussa Maazou, Milan Jovanović, Ognjen Vukojević and Rob Rensenbrink.

Because of the physical nature of Belgian football, it has tended to primarily produce talented defensive players. These include Jean-Marie Pfaff, Eric Gerets, Leo Clijsters, Michel Preud'homme, Georges Grün, Philippe Albert, Franky Van Der Elst, Vincent Kompany and Thomas Vermaelen. In comparison, only few attacking Belgian footballers have received international recognition: Enzo Scifo, Jan Ceulemans, Marc Degryse, Luc Nilis, Emile Mpenza, Romelu Lukaku, Eden Hazard etc.

Clubs

Matricule numbers

With football's rapid growth in popularity in the late 19th century, several football clubs came into existence in Belgium. The first to register with the national association was R. Antwerp F.C.. They were subsequently assigned the matricule number 1 when the Royal Belgian Football Association gave a matricule number to each registered club in November 1926, by order of registration. Many matricule numbers no longer exist due to clubs ceasing to exist or merging with another club. When two (or more) clubs merge, they must choose which matricule number to keep. The new club begins the championship at the level where the former club with the same matricule number should have begun the season. Typically they choose the one with the most honours. However, it has often occurred that a club with multiple championship titles have to merge with another less successful club in order to survive. In this case, the club with financial problems has to take the matricule number of the other club, and the honours are lost with the merger. For instance, when the 7 times champion Antwerp-based club K Beerschot VAC was struggling with financial difficulties in the third division in the late 1990s, they merged with then first division neighbour club KFC Germinal Ekeren to survive. The new club named KFC Germinal Beerschot Antwerpen started in the first division with the matricule number of KFC Germinal Ekeren, losing the honours of K Beerschot VAC, but keeping their installations and their purple shirt. Another famous example is that of 5 times champion Daring Club de Bruxelles merger with RR White into R White Daring Molenbeek in 1973.

The first few matricule numbers are:

1. R. Antwerp F.C..
2. Daring Club de Bruxelles (no longer active)
3. Club Brugge
4. R.F.C. Liégeois
5. R. Léopold Uccle Forestoise
6. Racing Club de Bruxelles
7. K.A.A. Gent
8. R.C.S. Verviétois
9. R. Dolhain F.C.
10. R. Union Saint-Gilloise

For a more complete list, see List of football clubs in Belgium.

Naming

A Belgian club's name usually includes the name of the town where the club plays as well as a prefix and/or suffix. Since Belgians speak three languages, French and Dutch being the main ones and German being the third official language, Belgian teams may use either language as the basis for their names. For historical reasons, many Flemish clubs changed their names from French to Dutch between the beginning of the 20th century and the late 1960s. Additionally, many clubs have experienced frequent name changes. Reasons for these include a language change, a merger, an anniversary, etc. Because of the numerous mergers between Belgian clubs, team names sometimes combine several town names (*such as* K. Beringen-Heusden-Zolder *or* Sporting West Ingelmunster-Harelbeke) which reflect mergers. In recent history clubs representing immigrant communities have come into existence and sometimes use names that are in neither of Belgium's official languages (the now defunct clubs Türkgucun Ozburun and Türkiyemspor Zaventem, or the still existing Agrupacion Oviedo-Asturiana from Brussels being examples).

Finally, a team which exists for at least 50 years may add the prefix *Royal* to its name (either in English or in the team's language). Before 1958, this right was given to any team that celebrated its 25th year of existence. Between 1958 and 1968, the rule was changed to grant the title to any team with at least 35 years of existence. Since 1968, the time limit has increased to 50 years.

The following is a partial list of common prefixes and suffixes in Belgium's two main languages.

Name	French	Dutch
Royal	Royale (R.)	Koninklijke (K.)
Athletic Association	Association Athlétique (A.A.)	Atletieke Associatie (A.A.)
Sport Association	Association Sportive (A.S.)	Sportvereniging (S.V.)
Sport Circle	Cercle Sportif (C.S.)	Sportkring (S.K.)
Excelsior	Excelsior (E.)	Excelsior (E.)
Football Club	Football Club (F.C.)	Voetbalklub (V.K.) or Voetbalclub (V.C.)
Sport Club	Sporting Club (S.C.)	Sporting Club or Sport Club (S.C.)
Racing Club	Racing Club (R.C.)	Racing Club (R.C.)
Football Association	Association Football (A.F.)	Voetbal Vereniging (V.V.)
Royal Society	Société Royale (S.R.)	Koninklijke Maatschappij (K.M.)
Athletic Club	Athletic Club (A.C.)	Atletiek Club (A.C.)

European results

R.S.C. Anderlecht and Y.R. K.V. Mechelen have won a European competition. Here is a list of the winners and runners-up by competition:

- UEFA Champions' League:
  - Final:
    - Club Brugge (1978)
- Cup Winners' Cup:
  - Wins:
    - R.S.C. Anderlecht (1976 and 1978)
    - K.V. Mechelen (1988)
  - Finals:
    - R.S.C. Anderlecht (1977 and 1990)
    - Standard Liège (1982)
    - R. Antwerp F.C. (1993)
- UEFA Cup:
  - Wins:
    - R.S.C. Anderlecht (1983)
  - Finals:
    - Club Brugge (1976)
    - R.S.C. Anderlecht (1984)
- Inter-Cities Fairs Cup:
  - Final:
    - R.S.C. Anderlecht (1970)
- European Supercup:
  - Wins:
    - R.S.C. Anderlecht (1976 and 1978)

- K.V. Mechelen (1988)

Football in Belgium by province

Under the first four levels in the league system, the competition is organized by province, with the notable exception of the Brabant that comprises clubs from the provinces of Flemish Brabant, Walloon Brabant and the Brussels Capital Region.

West Flanders

Four clubs from West Flanders play in the first division (Cercle and Club Brugge, KV Kortrijk and Zulte-Waregem) and two clubs play in the second division (Oostende and K.S.V. Roeselare).
Bruges is the only city in Belgium with two teams in the first division.

East Flanders

Two clubs from East Flanders play in the first division (Gent and Lokeren).

Antwerp

The province of Antwerp has an old tradition of football. The first Belgian clubs were created in the city of Antwerp (Antwerp Lyon's Club, A.S. Anvers-Borgerhout, Antwerp F.C.). Four clubs from this province play at the highest level (Germinal Beerschot, Mechelen, Lierse and Westerlo).

Limburg

Two clubs from Limburg play in the Jupiler League (Genk and Sint-Truidense).

Brabant

Only one club from Brabant played in the first division (Anderlecht) during several years. After the season 2010-2011 Oud-Heverlee Leuven also came in.

The first Belgian championship ever was played between 7 teams, 4 of which were based in Brussels: Racing Club, Léopold Club, Sporting Club and Union d'Ixelles. Léopold Club was a club for the nobility and bourgeoisie in Brussels and is still alive following a string of 4 merges between 1982 and 2001 (but it plays in the Promotion B). The latter two clubs collapsed respectively in 1897 and 1901. That year a new club from Brussels arrived in the first division (namely R. Union Saint-Gilloise) and will become a leading team in Belgian football in the following seasons. It enjoyed first a rivalry with Racing Club and then with Daring Club (that won its first championship) in 1912. Later, they played alongside Anderlecht in the first division.

After World War I, Belgian football was dominated by teams from the Antwerp province and from Brugge. In the early 1930s, Union and Daring shared 5 titles. The rivalry between the two clubs has inspired a play named *Bossemans et Coppenolle* (Bossemans was the name of the Union coach and Coppenolle was the name of the Daring coach). During World War II, Anderlecht replaced Union and Daring as the leading team in Brussels. Its rivals in Brussels were successively Union, Daring, Racing White and its successors RWDM and FC Brussels.

Hainaut

Hainaut only has one team that will be playing in 2011 in the Belgian Pro League: R.A.E.C Mons. There will be a few teams from Hainaut playing in the Belgian 2nd Division (Exqi League) in 2011 though.

Namur

The province of Namur is, with that of Luxembourg, the province with the least prestigious football history in Belgium. No club from this province is playing in one of the first two levels in Belgian football. No club from the province of Namur has ever played in the first division.

Liège

The first Belgian championship ever was won by FC Liégeois. In 2011, only one club from the province of Liège will be playing in the first division (Standard Liège). K.A.S.Eupen was the first and so far only club ever from the German speaking area of Belgium to reach the top division. Their official name is: Königliche Allgemeine Sportvereinigung Eupen.

Football in the province of Liège was first led by FC Liégeois (5 titles from 1895 to 1953) and then by Standard Liège (9 titles from 1958 to 2008). CS Verviers played some seasons in the first division, as well as FC Sérésien (that later merged with Standard) and FC Tilleur (that later associated, and not merged, with FC Liégeois).

Luxembourg

Only one club from the province of Luxembourg used to play in the second division: Virton. Philippe Albert was born in this province (in the municipality of Bouillon). No club from this province has ever played in the first division.

League system

There are 8 level leagues in Belgian football.

European Competitions

UEFA Champions League

The following teams have qualified for group stage in the UEFA Champions League

- Club Brugge K.V. (4 times: 1992-93, 2002-03, 2003-04, 2005-06)
- Anderlecht (8 times: 1993-94, 1994-95, 2000–01, 2001–02, 2003-04, 2004-05, 2005-06, 2006-07)
- K. Lierse S.K. (1997–98)
- Genk (2 times: 2002-03, 2011-12)
- Standard Liège (2009-10)

Football outside of the Belgian FA

Other than the UEFA-affiliated FA, several amateur football leagues exist in Belgium, most at regional levels. These are often called "pub teams' leagues" while this is not entirely correct: far from all clubs represent a pub, often there are teams enroled who represent a town or district of a village, a company, or another institution. Many amateur leagues exist in Belgium, most of them are region-bound or province-bound. Examples of amateur leagues with a long tradition include the KVV (Koninklijke Vlaamse Voetbalbond) organising provincial leagues in all Flemish provinces but West-Vlaanderen, the MTSA for the area Dendermonde-Aalst-Denderstreek, the LVVB Melle for the area Wetteren-Ghent (with some clubs outside of this area), the LVV Meetjesland for the Meetjesland area and few clubs from Ghent, the Corporative Leagues in several provinces mainly intended for company teams (although sometimes also including general amateur teams), the WALIVO in the Waasland area, the ABSSA and Travailliste leagues in Brussels and surrounding areas, ... The system is complex as some of these regional leagues are affiliated to the Belgian FA (sometimes their member clubs receive a matricule number) but the leagues are totally separate from the league system in the leagues directly run by the Belgian FA (described above). Some other amateur leagues operate totally separate from the Belgian FA with no connection to the Belgian FA in any ways. Amateur leagues are in decline in some areas due to the competition from the general Belgian FA-run leagues and due to long travel distances being unpractical at amateur level. Several clubs who started in amateur leagues have made the transfer to the leagues run directly by the Belgian FA (starting at the lowest level). Some of these clubs even managed to, after a while, reach a relatively high level of the Belgian pyramid. Many amateur leagues in the past were tied to a political ideology, and Catholic, Socialist and Liberal amateur leagues existed. Nowadays most amateur leagues are based upon geographic area rather than on political ideologies.

Indoor football

In Belgium, the Belgian FA also runs a nationwide Futsal league. Clubs are given a matricule number as well, although a separate one than the matricule numbers assigned to the clubs in outdoor football. Usually the letter A is in front of the matricule number to indicate an indoor club. A separate indoor football league is organised by the BZVB (Belgische Zaalvoetbal Bond, translated as Belgian Indoor Football Association), this is not tied to the Belgian FA. Both of these leagues run at the same moment but without any interaction with each other.

In addition, Mini Football is popular in Belgium (this has different rules all together than Futsal) and this league is run by a separate FA dealing with mini football only.

Women's football

Both the Belgian FA and several regional amateur football leagues run a league for women. The league operated by the Belgian FA consists of 3 nationwide levels, with several levels per province below. The Belgian FA started to organise women's football in the early seventies, due to the UEFA obligating every member FA to organise leagues for women as well as for men. In the beginning the Belgian FA disencouraged women to play football and advised them to opt for sports such as volley ball. The first season only existing clubs (with male teams) could enrol a female team in the league. As popularity of women's football grew and more and more teams wanted to play in the league, the Belgian FA dropped the above rule and accepted new clubs to affiliate who only focussed on ladies' football. These clubs are assigned matricule numbers just like any other club, and meanwhile women's football is fully integrated in the Belgian football. That said, the league is not professional as yet (only a few female players have been full-time professionals) and the national team is amongst the weaker teams in Europe due to other countries such as Germany, Norway, Sweden having fully professional women's leagues. Clubs who have been successful in past or present in Belgian women's football include Brussels Dames '71 (currently the women's team of RSC Anderlecht), Rapide Wezemaal, Astro Begijnendijk, Eva's Kumtich, Sinaai Girls, Standard Fémina Liège, Dames Eendracht Aalst (previously tied to the club KSC Eendracht Aalst which was also successful in men's football). Plans exist to develop a new super league where existing clubs with professional men's teams would enrole a women's team. While this new league is not launched yet, several famous clubs have already created a women's team, such as Club Brugge, Lierse SK, Germinal Beerschot Antwerpen, and St-Truidense VV.

References

- (English) RSSSF archive [1] - Belgian clubs history
- (Dutch) Belgian Soccer Database [2] - Extensive database about players, clubs, results and other historical statistics

References

[1] http://users.skynet.be/pluto/Textbelgium/Teams.html
[2] http://www.bsdb.be/

Association_football

An attacking player (No. 10) attempts to kick the ball past the opposing team's goalkeeper and between the goalposts to score a goal.

Highest governing body	FIFA
Nickname(s)	Football, soccer, footy/footie, "the beautiful game", "the world game"
First played	Mid-19th century Great Britain, Nottinghamshire, Sheffield.
Characteristics	
Contact	Yes
Team members	11 per side
Mixed gender	Yes, separate competitions
Categorization	Team sport, ball sport
Equipment	Football (or soccer ball)
Venue	Football pitch (or soccer field)
Olympic	Part of the Summer Olympic programme from 1900 to 1928 Returned to the Summer Olympic programme in 1936 to today
Paralympic	No
Country or region	Worldwide

Association football, more commonly known as **football** or **soccer**, is a sport played between two teams of eleven players with a spherical ball. At the turn of the 21st century, the game was played by over 250 million players in over 200 countries, making it the world's most popular sport.[1] [2] [3] [4] The game is played on a rectangular field of grass or green artificial turf, with a goal in the middle of each of the short ends. The object of the game is to score by driving the ball into the opposing goal.

In general play, the goalkeepers are the only players allowed to touch the ball with their hands or arms (unless the ball is carried out of play, where the field players are required to re-start by a throw-in of the game ball), while the field players typically use their feet to kick the ball into position, occasionally using their torso or head to intercept a ball in midair. The team that scores the most goals by the end of the match wins. If the score is tied at the end of the game, either a draw is declared or the game goes into extra time and/or a penalty shootout, depending on the format of the competition. The Laws of the Game were originally codified in England by the Football Association in 1863 and have evolved since then. Association football is governed internationally by FIFA - Fédération Internationale de Football Association (English: International Federation of Association Football), which organises the FIFA World Cup every four years.[5]

Etymology and names

The rules of football were codified in England by the Football Association in 1863 and the name *association football* was coined to distinguish the game from the other forms of football played at the time, specifically rugby football. The term *soccer* originated in England, first appearing in the 1880s as an Oxford "-er" abbreviation of the word "association".[6]

Within the English-speaking world, association football is usually called *football* (colloquially *footy*) or sometimes *soccer* in the United Kingdom, and mainly *soccer* in Australia, Canada, New Zealand, and the United States. Other countries may use either or both terms, and may also have local names for the sport.

Gameplay

A goalkeeper saving a close-range shot from inside the penalty area

Association football is played in accordance with a set of rules known as the Laws of the Game. The game is played using a spherical ball (of **unknown operator: u'strong'** cm (28 in) circumference in FIFA play), known as the *football* (or *soccer ball*). Two teams of eleven players each compete to get the ball into the other team's goal (between the posts and under the bar), thereby scoring a goal. The team that has scored more goals at the end of the game is the winner; if both teams have scored an equal number of goals then the game is a draw. Each team is led by a captain who has only one official responsibility as mandated by the Laws of the Game: to be involved in the coin toss prior to kick-off or penalty kicks.[7]

The primary law is that players other than goalkeepers may not deliberately handle the ball with their hands or arms during play, though they do use their hands during a throw-in restart. Although players usually use their feet to move the ball around, they may use any part of their body (notably, "heading" with the forehead)[8] other than their hands or arms.[9] Within normal play, all players are free to play the ball in any direction and move throughout the pitch, though the ball cannot be received in an offside position.[10]

In typical game play, players attempt to create goal-scoring opportunities through individual control of the ball, such as by dribbling, passing the ball to a team-mate, and by taking shots at the goal, which is guarded by the opposing goalkeeper. Opposing players may try to regain control of the ball by intercepting a pass or through tackling the opponent in possession of the ball; however, physical contact between opponents is restricted. Football is generally a free-flowing game, with play stopping only when the ball has left the field of play or when play is stopped by the referee for an infringement of the rules. After a stoppage, play recommences with a specified restart.[11]

A goalkeeper dives to stop the ball from entering his goal

At a professional level, most matches produce only a few goals. For example, the 2005–06 season of the English Premier League produced an average of 2.48 goals per match.[12] The Laws of the Game do not specify any player positions other than goalkeeper,[13] but a number of specialised roles have evolved. Broadly, these include three main categories: strikers, or forwards, whose main task is to score goals; defenders, who specialise in preventing their opponents from scoring; and midfielders, who dispossess the opposition and keep possession of the ball in order to pass it to the forwards on their team. Players in these positions are referred to as outfield players, in order to distinguish them from the goalkeeper. These positions are further subdivided according to the area of the field in which the player spends most time. For example, there are central defenders, and left and right midfielders. The ten outfield players may be arranged in any combination. The number of players in each position determines the style of the team's play; more forwards and fewer defenders creates a more aggressive and offensive-minded game, while the reverse creates a slower, more defensive style of play. While players typically spend most of the game in a specific position, there are few restrictions on player movement, and players can switch positions at any time.[14] The layout of a team's players is known as a *formation*. Defining the team's formation and tactics is usually the prerogative of the team's manager.[15]

History

England playing Scotland in the first-ever international football game (The Oval, 1872)

Games revolving around the kicking of a ball have been played in many countries throughout history. According to FIFA, the "very earliest form of the game for which there is scientific evidence was an exercise from a military manual dating back to the second and third centuries BC in China", which was known as *cuju*.[16] The modern rules of association football are based on the mid-19th century efforts to standardise the widely varying forms of football played at the public schools of England. The history of football in England dates back to at least the eighth century.[17]

The Royal Engineers team who reached the first FA Cup final in 1872

The Cambridge Rules, first drawn up at Cambridge University in 1848, were particularly influential in the development of subsequent codes, including association football. The Cambridge Rules were written at Trinity College, Cambridge, at a meeting attended by representatives from Eton, Harrow, Rugby, Winchester and Shrewsbury schools. They were not universally adopted. During the 1850s, many clubs unconnected to schools or universities were formed throughout the English-speaking world, to play various forms of football. Some came up with their own distinct codes of rules, most notably the Sheffield Football Club, formed by former public school pupils in 1857,[18] which led to formation of a Sheffield FA in 1867. In 1862, John Charles Thring of Uppingham School also devised an influential set of rules.[19]

These ongoing efforts contributed to the formation of The Football Association (The FA) in 1863, which first met on the morning of 26 October 1863 at the Freemasons' Tavern in Great Queen Street, London.[20] The only school to be represented on this occasion was Charterhouse. The Freemason's Tavern was the setting for five more meetings between October and December, which eventually produced the first comprehensive set of rules. At the final meeting, the first FA treasurer, the representative from Blackheath, withdrew his club from the FA over the removal of two draft rules at the previous meeting: the first allowed for running with the ball in hand; the second for obstructing such a run by hacking (kicking an opponent in the shins), tripping and holding. Other English rugby football clubs followed this lead and did not join the FA, or subsequently left the FA and instead in 1871 formed the Rugby Football Union. The eleven remaining clubs, under the charge of Ebenezer Cobb Morley, went on to ratify the original thirteen laws of the game.[20] These rules included handling of the ball by "marks" and the lack of a crossbar, rules which made it remarkably similar to Victorian rules football being developed at that time in Australia. The Sheffield FA played by its own rules until the 1870s with the FA absorbing some of its rules until there was little difference between the games.[21]

The laws of the game are currently determined by the International Football Association Board (IFAB).[22] The Board was formed in 1886[23] after a meeting in Manchester of The Football Association, the Scottish Football Association, the Football Association of Wales, and the Irish Football Association. The world's oldest football competition is the FA Cup, which was founded by C. W. Alcock and has been contested by English teams since 1872. The first official international football match took place in 1872 between Scotland and England in Glasgow, again at the instigation of C. W. Alcock. England is home to the world's first football league, which was founded in Birmingham in 1888 by Aston Villa director William McGregor.[24] The original format contained 12 clubs from the Midlands and the North of England. FIFA, the international football body, was formed in Paris in 1904 and declared that they would adhere to Laws of the Game of the Football Association.[25] The growing popularity of the international game led to the admittance of FIFA representatives to the International Football Association Board in 1913. The board currently consists of four representatives from FIFA and one representative from each of the four British associations.[26]

Today, football is played at a professional level all over the world. Millions of people regularly go to football stadiums to follow their favourite teams,[27] while billions more watch the game on television or on the internet.[28] A very large number of people also play football at an amateur level. According to a survey conducted by FIFA published in 2001, over 240 million people from more than 200 countries regularly play football.[29] Football has the highest global television audience in sport.[30]

In many parts of the world football evokes great passions and plays an important role in the life of individual fans, local communities, and even nations. R. Kapuscinski says that people who are polite, modest or even humble in Europe fall easily into rage with playing or watching soccer games.[31] The Côte d'Ivoire national football team helped secure a truce to the nation's civil war in 2006[32] and it helped further reduce tensions between government and rebel forces in 2007 by playing a match in the rebel capital of Bouaké, an occasion that brought both armies together peacefully for the first time.[33] By contrast, football is widely considered to be the final proximate cause in the Football War in June 1969 between El Salvador and Honduras.[34] The sport also exacerbated tensions at the beginning of the Yugoslav wars of the 1990s, when a match between Dinamo Zagreb and Red Star Belgrade degenerated into rioting in May 1990.[35]

Laws

There are 17 laws in the official Laws of the Game, each containing a collection of stipulation and guidelines. The same laws are designed to apply to all levels of football, although certain modifications for groups such as juniors, seniors, women and people with physical disabilities are permitted. The laws are often framed in broad terms, which allow flexibility in their application depending on the nature of the game. The Laws of the Game are published by FIFA, but are maintained by the International Football Association Board (IFAB).[36] In addition to the seventeen laws, numerous IFAB decisions and other directives contribute to the regulation of football.

Players, equipment, and officials

Each team consists of a maximum of eleven players (excluding substitutes), one of whom must be the goalkeeper. Competition rules may state a minimum number of players required to constitute a team, which is usually seven. Goalkeepers are the only players allowed to play the ball with their hands or arms, provided they do so within the penalty area in front of their own goal. Though there are a variety of positions in which the outfield (non-goalkeeper) players are strategically placed by a coach, these positions are not defined or required by the Laws.[13]

The basic equipment or *kit* players are required to wear includes a shirt, shorts, socks, footwear and adequate shin guards. Headgear is not a required piece of basic equipment, but players today may choose to wear it to protect themselves from head injury. Players are forbidden to wear or use anything that is dangerous to themselves or another player, such as jewellery or watches. The goalkeeper must wear clothing that is easily distinguishable from that worn by the other players and the match officials.[37]

A number of players may be replaced by substitutes during the course of the game. The maximum number of substitutions permitted in most competitive international and domestic league games is three, though the permitted number may vary in other competitions or in friendly matches. Common reasons for a substitution include injury, tiredness, ineffectiveness, a tactical switch, or timewasting at the end of a finely poised game. In standard adult matches, a player who has been substituted may not take further part in a match.[38] IFAB recommends that "that a match should not continue if there are fewer than seven players in either team." Any decision regarding points awarded for abandoned games is left to the individual football associations.[39]

A game is officiated by a referee, who has "full authority to enforce the Laws of the Game in connection with the match to which he has been appointed" (Law 5), and whose decisions are final. The referee is assisted by two assistant referees. In many high-level games there is also a fourth official who assists the referee and may replace another official should the need arise.[40]

Pitch

As the Laws were formulated in England, and were initially administered solely by the four British football associations within IFAB, the standard dimensions of a football pitch were originally expressed in imperial units. The Laws now express dimensions with approximate metric equivalents (followed by traditional units in brackets), though popular use tends to continue to use traditional units in English-speaking countries with a relatively recent history of metrication (or only partial metrication), such as Britain.[41]

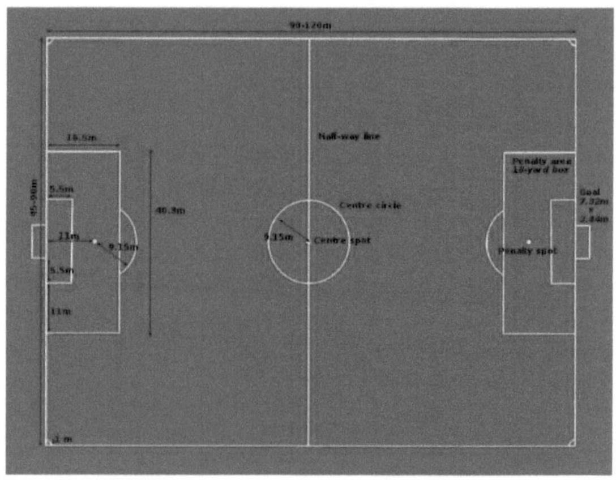

Standard pitch measurements (See Imperial version)

The length of the pitch for international adult matches is in the range of 100–110 m (110–120 yd) and the width is in the range of 64–75 m (70–80 yd). Fields for non-international matches may be 90–120 m (100–130 yd) length and 45–90 m (50–100 yd) in width, provided that the pitch does not become square. Although in 2008, the IFAB initially approved a fixed size of 105 m (344 ft) long and 68 m (223 ft) wide as a standard pitch dimension for A international matches,[42] this decision was later put on hold and was never actually implemented.[43]

The longer boundary lines are *touchlines*, while the shorter boundaries (on which the goals are placed) are *goal lines*. A rectangular goal is positioned at the middle of each goal line.[44] The inner edges of the vertical goal posts must be 7.32 m (8 yd) apart, and the lower edge of the horizontal crossbar supported by the goal posts must be 2.44 m (8 ft) above the ground. Nets are usually placed behind the goal, but are not required by the Laws.[45]

In front of each goal is an area known as the penalty area. This area is marked by the goal line, two lines starting on the goal line 16.5 m (18 yd) from the goalposts and extending 16.5 m (18 yd) into the pitch perpendicular to the goal line, and a line joining them. This area has a number of functions, the most prominent being to mark where the goalkeeper may handle the ball and where a penalty foul by a member of the defending team becomes punishable by a penalty kick. Other markings define the position of the ball or players at kick-offs, goal kicks, penalty kicks and corner kicks.[46]

Duration and tie-breaking methods

A standard adult football match consists of two periods of 45 minutes each, known as halves. Each half runs continuously, meaning that the clock is not stopped when the ball is out of play. There is usually a 15-minute half-time break between halves. The end of the match is known as full-time.[47] The referee is the official timekeeper for the match, and may make an allowance for time lost through substitutions, injured players requiring attention, or other stoppages. This added time is most commonly referred to as *stoppage time* or *injury time*, while *loss time* can also be used as a synonym. The duration of stoppage time is at the sole discretion of the referee. The referee alone signals the end of the match. In matches where a fourth official is appointed, toward the end of the half the referee signals how many minutes of stoppage time he intends to add. The fourth official then informs the players and spectators by holding up a board showing this number. The signalled stoppage time may be further extended by the referee.[47] Added time was introduced because of an incident which happened in 1891 during a match between Stoke and Aston Villa. Trailing 1–0 and with just two minutes remaining, Stoke were awarded a penalty. Villa's goalkeeper kicked the ball out of the ground, and by the time the ball had been recovered, the 90 minutes had elapsed and the game was over.[48] The same law also stands that the duration of either half is extended until the penalty kick to be taken or retaken is completed, thus no game shall end with a penalty to be taken.[49]

In league competitions, games may end in a draw. In knockout competitions where a winner is required various methods may be employed to break such a deadlock, some competitions may invoke replays.[50] A game tied at the end of regulation time it may go into extra time, which consists of two further 15-minute periods. If the score is still tied after extra time, some competitions allow the use of penalty shootouts (known officially in the Laws of the Game as "kicks from the penalty mark") to determine which team will progress to the next stage of the tournament. Goals scored during extra time periods count toward the final score of the game, but kicks from the penalty mark are only used to decide the team that progresses to the next part of the tournament (with goals scored in a penalty shootout not making up part of the final score).[7]

In competitions using two-legged matches, each team competes at home once, with an aggregate score from the two matches deciding which team progresses. Where aggregates are equal, the away goals rule may be used to determine the winners, in which case the winner is the team that scored the most goals in the leg they played away from home. If the result is still equal, extra time and potentially a penalty shootout are required.[7]

In the late 1990s and early 2000s, the IFAB experimented with ways of creating a winner without requiring a penalty shootout, which was often seen as an undesirable way to end a match. These involved rules ending a game in extra time early, either when the first goal in extra time was scored (*golden goal*), or if one team held a lead at the end of the first period of extra time (*silver goal*). Golden goal was used at the World Cup in 1998 and 2002. The first World Cup game decided by a golden goal was France's victory over Paraguay in 1998. Germany was the first nation to score a golden goal in a major competition, beating Czech Republic in the final of Euro 1996. Silver goal was used in Euro 2004. Both these experiments have been discontinued by IFAB.[51]

Ball in and out of play

Under the Laws, the two basic states of play during a game are *ball in play* and *ball out of play*. From the beginning of each playing period with a kick-off until the end of the playing period, the ball is in play at all times, except when either the ball leaves the field of play, or play is stopped by the referee. When the ball becomes out of play, play is restarted by one of eight restart methods depending on how it went out of play:

- Kick-off: following a goal by the opposing team, or to begin each period of play.[11]
- Throw-in: when the ball has crossed the touchline; awarded to opposing team to that which last touched the ball.[52]
- Goal kick: when the ball has wholly crossed the goal line without a goal having been scored and having last been touched by a player of the attacking team; awarded to defending team.[53]
- Corner kick: when the ball has wholly crossed the goal line without a goal having been scored and having last been touched by a player of the defending team; awarded to attacking team.[54]

A player takes a free kick, while the opposition form a "wall" to try to block the ball

- Indirect free kick: awarded to the opposing team following "non-penal" fouls, certain technical infringements, or when play is stopped to caution or send-off an opponent without a specific foul having occurred. A goal may not be scored directly (without the ball first touching another player) from an indirect free kick.[55]
- Direct free kick: awarded to fouled team following certain listed "penal" fouls.[55] A goal may be scored directly from a direct free kick.
- Penalty kick: awarded to the fouled team following a foul usually punishable by a direct free kick but that has occurred within their opponent's penalty area.[56]
- Dropped-ball: occurs when the referee has stopped play for any other reason, such as a serious injury to a player, interference by an external party, or a ball becoming defective.[11]

Misconduct

Players are cautioned with a yellow card, and sent off with a red card. These colours were first introduced at the 1970 FIFA World Cup and used consistently since.

A player scores a penalty kick given after an offence is committed inside the penalty area

A foul occurs when a player commits an offence listed in the Laws of the Game while the ball is in play. The offences that constitute a foul are listed in Law 12. Handling the ball deliberately, tripping an opponent, or pushing an opponent, are examples of "penal fouls", punishable by a direct free kick or penalty kick depending on where the offence occurred. Other fouls are punishable by an indirect free kick.[9] The referee may punish a player or substitute's misconduct by a caution (yellow card) or sending-off (red card). A second yellow card at the same game leads to a red card, and therefore to a sending-off. A player given a yellow card is said to have been "booked", the referee writing the player's name in his official notebook. If a player has been sent off, no substitute can be brought on in their place. Misconduct may occur at any time, and while the offences that constitute misconduct are listed, the definitions are broad. In particular, the offence of "unsporting behaviour" may be used to deal with most events that violate the spirit of the game, even if they are not listed as specific offences. A referee can show a yellow or red card to a player, substitute or substituted player. Non-players such as managers and support staff cannot be shown the yellow or red card, but may be expelled from the technical area if they fail to conduct themselves in a responsible manner.[9]

Rather than stopping play, the referee may allow play to continue if doing so will benefit the team against which an offence has been committed. This is known as "playing an advantage".[57] The referee may "call back" play and penalise the original offence if the anticipated advantage does not ensue within "a few seconds". Even if an offence is not penalised due to advantage being played, the offender may still be sanctioned for misconduct at the next stoppage of play.[58]

Governing bodies

The recognised international governing body of football (and associated games, such as futsal and beach soccer) is the Fédération Internationale de Football Association (FIFA). The FIFA headquarters are located in Zurich. Six regional confederations are associated with FIFA; these are:[59]

- Asia: Asian Football Confederation (AFC)
- Africa: Confederation of African Football (CAF)
- Europe: Union of European Football Associations (UEFA)
- North/Central America & Caribbean: Confederation of North, Central American and Caribbean Association Football (CONCACAF)
- Oceania: Oceania Football Confederation (OFC)
- South America: Confederación Sudamericana de Fútbol/Confederação Sul-americana de Futebol (South American Football Confederation; CONMEBOL)

National associations oversee football within individual countries. These are generally synonymous with sovereign states, (for example: the Fédération Camerounaise de Football in Cameroon) but also include a smaller number of associations responsible for sub-national entities or autonomous regions (for example the Scottish Football Association in Scotland). 208 national associations are affiliated both with FIFA and with their respective continental confederations.[59]

While FIFA is responsible for arranging competitions and most rules related to international competition, the actual Laws of the Game are set by the International Football Association Board, where each of the UK Associations has one vote, while FIFA collectively has four votes.[26]

International competitions

The major international competition in football is the World Cup, organised by FIFA. This competition takes place over a four-year period. More than 190 national teams compete in qualifying tournaments within the scope of continental confederations for a place in the finals. The finals tournament, which is held every four years, involves 32 national teams competing over a four-week period.[60] The most recent tournament, the 2010 FIFA World Cup, was held in South Africa from 11 June to 11 July.[61]

A minute's silence before an international match

There has been a football tournament at every Summer Olympic Games since 1900, except at the 1932 games in Los Angeles.[62] Before the inception of the World Cup, the Olympics (especially during the 1920s) had the same status as the World Cup. Originally, the event was for amateurs only,[25] however, since the 1984 Summer Olympics professional players have been permitted, albeit with certain restrictions which prevent countries from fielding their strongest sides. Currently, the Olympic men's tournament is played at Under-23 level. In the past the Olympics have allowed a restricted number of over-age players per team;[63] but that practice ceased in the 2008 Olympics. A women's tournament was added in 1996; in contrast to the men's event, full international sides without age restrictions play the women's Olympic tournament.[64]

After the World Cup, the most important international football competitions are the continental championships, which are organised by each continental confederation and contested between national teams. These are the European Championship (UEFA), the Copa América (CONMEBOL), African Cup of Nations (CAF), the Asian Cup (AFC), the CONCACAF Gold Cup (CONCACAF) and the OFC Nations Cup (OFC). The FIFA Confederations Cup is contested by the winners of all 6 continental championships, the current FIFA World Cup champions and the country which is hosting the Confederations Cup. This is generally regarded as a warm-up tournament for the upcoming FIFA World Cup and does not carry the same prestige as the World Cup itself. The most prestigious competitions in club football are the respective continental championships, which are generally contested between national champions, for example the UEFA Champions League in Europe and the Copa Libertadores de América in South America. The winners of each continental competition contest the FIFA Club World Cup.[65]

Domestic competitions

The governing bodies in each country operate league systems in a domestic season, normally comprising several divisions, in which the teams gain points throughout the season depending on results. Teams are placed into tables, placing them in order according to points accrued. Most commonly, each team plays every other team in its league at home and away in each season, in a round-robin tournament. At the end of a season, the top team is declared the champion. The top few teams may be promoted to a higher division, and one or more of the teams finishing at the bottom are relegated to a lower division.[66] The teams finishing at the top of a country's league may be eligible also to play in international club competitions in the following season. The main exceptions to this system occur in some Latin American leagues, which divide football championships into two sections named Apertura and Clausura (Spanish for *Opening* and *Closing*), awarding a champion for each.[67] The majority of countries supplement the league system with one or more "cup" competitions organised on a knock-out basis.

Some countries' top divisions feature highly paid star players; in smaller countries and lower divisions, players may be part-timers with a second job, or amateurs. The five top European leagues – the Premier League (England),[68] La Liga (Spain), Serie A (Italy), the Bundesliga (Germany) and Ligue 1 (France) – attract most of the world's best players and each of the leagues has a total wage cost in excess of £600 million/€763 million/US$1.185 billion.[69]

Women's association football

Two players trying to win the ball

Women have been playing association football since the first recorded women's game in 1895 in North London. It has traditionally been associated with charity games and physical exercise, particularly in the United Kingdom.[70] This perception began to change in the 1970s with the breakthrough of organised women's association football. Association football is the most prominent team sport for women in several countries, and one of the few women's team sports with professional leagues.

The growth in women's football has seen major competitions being launched at both national and international level mirroring the male competitions. Women's football faced many struggles throughout its fight for right. It had a "golden age" in the United Kingdom in the early 1920s when crowds reached 50,000 at some matches;[71] this was stopped on 5 December 1921 when England's Football Association voted to ban the game from grounds used by its member clubs. The FA's ban was rescinded in December 1969 with UEFA voting to officially recognise women's football in 1971.[70] The FIFA Women's World Cup was inaugurated in 1991 and has been held every four years since.[72]

Variants and casual play

Variants of football have been codified for reduced-sized teams (i.e. Five-a-side football) play in non-field environments (i.e. Beach soccer, Indoor soccer & Futsal) and for teams with disabilities (i.e. Paralympic association football).

One of the attractions of association football is that a casual game can be played with only minimal equipment – a basic game can be played on almost any open area of reasonable size with just a ball and items to mark the positions of two sets of goalposts. Such games can often have team sizes that vary considerably from 11-a-side, use a limited and/or modified subset of the official rules, and are likely to be self-officiated by the players.

See also

- Association football culture
- Association football tactics and skills
- List of association football clubs
- List of women's association football clubs
- List of men's national association football teams
- List of top association football goal scorers
- List of top association football goal scorers by country
- Lists of association football players
- Paralympic association football
- Variants of association football

References

[1] "Overview of Soccer" (http://www.britannica.com/EBchecked/topic/550852/football). Encyclopædia Britannica. . Retrieved 4 June 2008.

[2] Guttman, Allen (1993). "The Diffusion of Sports and the Problem of Cultural Imperialism" (http://books.google.com/books?id=tQY5wxQDn5gC&pg=PA129&lpg=PA129&dq=world's+most+popular+team+sport&source=web&ots=6ns3wVUEGV&sig=SZPKYSDMJBrO1uV4mPxNbKyAuJY#PPA129,M1). In Eric Dunning, Joseph A. Maguire, Robert E. Pearton. *The Sports Process: A Comparative and Developmental Approach*. Champaign: Human Kinetics. p. 129. ISBN 0-88011-624-2. . Retrieved 26 January 2008. "the game is complex enough not to be invented independently by many preliterate cultures and yet simple enough to become the world's most popular team sport"

[3] Dunning, Eric (1999). "The development of soccer as a world game" (http://books.google.com/books?id=X3lX_LVBaToC&pg=PA105&lpg=PA105&dq=world's+most+popular+team+sport&source=web&ots=ehee9Lr9o1&sig=nyvDhcrPoR8lXhYKE7k4CZYg_qU#PPA103,M1). *Sport Matters: Sociological Studies of Sport, Violence and Civilisation*. London: Routledge. p. 103. ISBN 0-415-06413-9. . Retrieved 26 January 2008. "During the twentieth century, soccer emerged as the world's most popular team sport"

[4] Mueller, Frederick; Cantu; Van Camp, Steven (1996). "Team Sports" (http://books.google.com/books?id=XG6AIHLtyaUC&pg=PA57&lpg=PA57&dq=soccer+most+popular+team+sport&source=web&ots=QzydYB5Am0&sig=w_ouIgmegjytYFfWy7k92guTNfU#PPA57,M1). *Catastrophic Injuries in High School and College Sports*. Champaign: Human Kinetics. p. 57. ISBN 0-87322-674-7. . Retrieved 26 January 2008. "Soccer is the most popular sport in the world and is an industry worth over US$400 billion world wide. 80% of this is generated in Europe, though its popularity is growing in the United States. It has been estimated that there were 22 million soccer players in the world in the early 1980s, and that number is increasing. In the United States soccer is now a major sport at both the high school and college levels"

[5] "2002 FIFA World Cup TV Coverage" (http://web.archive.org/web/20061230124633/http://www.fifa.com/en/marketing/newmedia/index/0,3509,10,00.html). FIFA. 5 December 2006. Archived from the original (http://www.fifa.com/en/marketing/newmedia/index/0,3509,10,00.html) on 30 December 2006. . Retrieved 6 January 2008.

[6] Mazumdar, Partha (5 June 2006). "The Yanks are Coming: A U.S. World Cup Preview" (http://www.usembassy.org.uk/rss/transcripts/worldcup2006a.html). Embassy of the United States in London. . Retrieved 6 June 2009.

[7] IFAB. "Procedures to determine the winner of a match or home-and-away" (http://www.fifa.com/mm/document/affederation/generic/81/42/36/lawsofthegame_2010_11_e.pdf) (PDF). *Laws of the Game 2010/2011*. FIFA. pp. 51–52. . Retrieved 4 March 2011.

[8] "How to head a football)" (http://expertfootball.com/training/heading.php). . Retrieved 3 January 2011.

[9] "Laws of the game (Law 12)" (http://web.archive.org/web/20071011115718/http://fifa.com/flash/lotg/football/en/Laws12_02.htm). FIFA. Archived from the original (http://www.fifa.com/flash/lotg/football/en/Laws12_02.htm) on 11 October 2007. . Retrieved 24 September 2007.

[10] IFAB. "Law 11 – Offside" (http://www.fifa.com/mm/document/affederation/generic/81/42/36/lawsofthegame_2010_11_e.pdf) (PDF). *Laws of the Game 2010/2011*. FIFA. p. 31. . Retrieved 4 March 2011.

[11] "Laws of the game (Law 8)" (http://web.archive.org/web/20070913142456/http://fifa.com/flash/lotg/football/en/Laws8_01.htm). FIFA. Archived from the original (http://www.fifa.com/flash/lotg/football/en/Laws8_01.htm) on 13 September 2007. . Retrieved 24 September 2007.

[12] "England Premiership (2005/2006)" (http://www.sportpress.com/stats/en/738_england_premiership_2005_2006/11_league_summary.html). *Sportpress.com*. . Retrieved 5 June 2007.

[13] "Laws of the game (Law 3–Number of Players)" (http://web.archive.org/web/20070913142527/http://fifa.com/flash/lotg/football/en/Laws3_01.htm). FIFA. Archived from the original (http://www.fifa.com/flash/lotg/football/en/Laws3_01.htm) on 13 September 2007. . Retrieved 24 September 2007.

[14] "Positions guide, Who is in a team?" (http://news.bbc.co.uk/sport1/hi/football/rules_and_equipment/4196830.stm). *BBC Sport* (BBC). 1 September 2005. . Retrieved 24 September 2007.

[15] "Formations" (http://news.bbc.co.uk/sport1/hi/football/rules_and_equipment/4197420.stm). *BBC Sport* (BBC). 1 September 2005. . Retrieved 24 September 2007.

[16] "History of Football" (http://www.fifa.com/classicfootball/history/game/historygame1.html). FIFA. . Retrieved 20 November 2006.

[17] "History of Football – Britain, the home of Football" (http://www.fifa.com/classicfootball/history/game/historygame2.html). FIFA. . Retrieved 20 November 2006.

[18] Harvey, Adrian (2005). *Football, the first hundred years*. London: Routledge. p. 126. ISBN 0-415-35018-2.

[19] Winner, David (28 March 2005). "The hands-off approach to a man's game" (http://www.timesonline.co.uk/article/0,,27-1544006,00.html). *The Times* (London). . Retrieved 7 October 2007.

[20] "History of the FA" (http://web.archive.org/web/20050407161619/http://www.thefa.com/TheFA/TheOrganisation/Postings/2004/03/HISTORY_OF_THE_FA.htm). Football Association (FA). Archived from the original (http://www.thefa.com/TheFA/TheOrganisation/Postings/2004/03/HISTORY_OF_THE_FA.htm) on 7 April 2005. . Retrieved 9 October 2007.

[21] Young, Percy M. (1964). *Football in Sheffield*. S. Paul. pp. 28–29.

[22] "IFAB" (http://www.fifa.com/aboutfifa/organisation/ifab/aboutifab.html). FIFA. . Retrieved 10 December 2011.

[23] "The International FA Board" (http://web.archive.org/web/20070422035010/http://access.fifa.com/en/history/history/0,3504,3,00.html). FIFA. Archived from the original (http://access.fifa.com/en/history/history/0,3504,3,00.html) on 22 April 2007. . Retrieved 2 September 2007.

[24] "The History Of The Football League" (http://www.football-league.co.uk/page/History/HistoryDetail/0,,10794~1357277,00.html). Football League. 22 September 2010. . Retrieved 4 March 2011.

[25] "Where it all began" (http://web.archive.org/web/20070608215029/http://access.fifa.com/en/history/history/0,3504,4,00.html). FIFA. Archived from the original (http://access.fifa.com/en/history/history/0,3504,4,00.html) on 8 June 2007. . Retrieved 8 June 2007.

[26] "The IFAB: How it works" (http://www.fifa.com/aboutfifa/organisation/ifab/aboutifab.html). FIFA. . Retrieved 4 March 2011.

[27] Ingle, Sean; Glendenning, Barry (9 October 2003). "Baseball or Football: which sport gets the higher attendance?" (http://football.guardian.co.uk/news/theknowledge/0,9204,1059366,00.html). *The Guardian* (UK). . Retrieved 5 June 2006.

[28] "TV Data" (http://web.archive.org/web/20070922225713/http://fifa.com/aboutfifa/marketingtv/factsfigures/tvdata.html). FIFA. Archived from the original (http://www.fifa.com/aboutfifa/marketingtv/factsfigures/tvdata.html) on 22 September 2007. . Retrieved 2 September 2007.

[29] "FIFA Survey: approximately 250 million footballers worldwide" (http://web.archive.org/web/20060915133001/http://access.fifa.com/infoplus/IP-199_01E_big-count.pdf) (PDF). FIFA. Archived from the original (http://access.fifa.com/infoplus/IP-199_01E_big-count.pdf) on 15 September 2006. . Retrieved 15 September 2006.

[30] "2006 FIFA World Cup broadcast wider, longer and farther than ever before" (http://www.fifa.com/aboutfifa/organisation/marketing/news/newsid=111247/). FIFA. 6 February 2007. . Retrieved 11 October 2009.

[31] Kapuscinski, Ryszard (2007). *The Soccer War*.

[32] Stormer, Neil (20 June 2006). "More than a game" (http://www.commongroundnews.org/article.php?sid=1&id=2079). *Common Ground News Service*. . Retrieved 2 March 2010.

[33] Austin, Merrill (10 July 2007). "Best Feet Forward" (http://www.vanityfair.com/culture/features/2007/07/ivorycoast200707). *Vanity Fair*. . Retrieved 2 March 2010.

[34] Dart, James; Bandini, Paolo (21 February 2007). "Has football ever started a war?" (http://football.guardian.co.uk/theknowledge/story/0,,2017161,00.html). *The Guardian* (London). . Retrieved 24 September 2007.

[35] Drezner, Daniel (4 June 2006). "The Soccer Wars" (http://www.washingtonpost.com/wp-dyn/content/article/2006/06/02/AR2006060201401.html). *The Washington Post*: p. B01. . Retrieved 21 May 2008.

[36] "Laws Of The Game" (http://www.fifa.com/worldfootball/lawsofthegame.html). FIFA. . Retrieved 2 September 2007.

[37] "Laws of the game (Law 4–Players' Equipment)" (http://web.archive.org/web/20070913141601/http://fifa.com/flash/lotg/football/en/Laws4_01.htm). FIFA. Archived from the original (http://www.fifa.com/flash/lotg/football/en/Laws4_01.htm) on 13 September 2007. . Retrieved 24 September 2007.

[38] "Laws of the game (Law 3–Substitution procedure)" (http://web.archive.org/web/20071011144947/http://fifa.com/flash/lotg/football/en/Laws3_02.htm). FIFA. Archived from the original (http://www.fifa.com/flash/lotg/football/en/Laws3_02.htm) on 11 October 2007. . Retrieved 24 September 2007.

[39] IFAB. "Law 3 – The Number of Players" (http://www.fifa.com/mm/document/affederation/generic/81/42/36/lawsofthegame_2010_11_e.pdf) (PDF). *Laws of the Game 2010/2011*. FIFA. p. 62. . Retrieved 4 March 2011.

[40] "Laws of the game (Law 5–The referee)" (http://web.archive.org/web/20070913141909/http://fifa.com/flash/lotg/football/en/Laws5_01.htm). FIFA. Archived from the original (http://www.fifa.com/flash/lotg/football/en/Laws5_01.htm) on 13 September 2007. . Retrieved 24 September 2007.

[41] Summers, Chris (2 September 2004). "Will we ever go completely metric?" (http://news.bbc.co.uk/1/hi/magazine/3934353.stm). *BBC news* (BBC). . Retrieved 7 October 2007.

[42] "Goal-line technology put on ice" (http://www.fifa.com/aboutfifa/organisation/ifab/media/news/newsid=707751/). FIFA. 8 March 2008. . Retrieved 19 June 2010.

[43] "FIFA Amendments to the Laws of the Game, 2008" (http://www.fifa.com/mm/document/affederation/administration/77/82/55/circularno.1145-amendmentstothelawsofthegame-2008.pdf) (PDF). FIFA. . Retrieved 4 March 2011.

[44] "Laws of the game (Law 1.1–The field of play)" (http://web.archive.org/web/20070913142202/http://fifa.com/flash/lotg/football/en/Laws1_01.htm). FIFA. Archived from the original (http://www.fifa.com/flash/lotg/football/en/Laws1_01.htm) on 13 September 2007. . Retrieved 24 September 2007.

[45] "Laws of the game (Law 1.4–The Field of play)" (http://web.archive.org/web/20071011144942/http://fifa.com/flash/lotg/football/en/Laws1_04.htm). FIFA. Archived from the original (http://www.fifa.com/flash/lotg/football/en/Laws1_04.htm) on 11 October 2007. . Retrieved 24 September 2007.

[46] "Laws of the game (Law 1.3–The field of play)" (http://web.archive.org/web/20071011084145/http://fifa.com/flash/lotg/football/en/Laws1_03.htm). FIFA. Archived from the original (http://www.fifa.com/flash/lotg/football/en/Laws1_03.htm) on 11 October 2007. . Retrieved 24 September 2007.

[47] "Laws of the game (Law 7.2–The duration of the match)" (http://web.archive.org/web/20071011144952/http://fifa.com/flash/lotg/football/en/Laws7_02.htm). FIFA. Archived from the original (http://www.fifa.com/flash/lotg/football/en/Laws7_02.htm) on 11 October 2007. . Retrieved 24 September 2007.

[48] The Sunday Times *Illustrated History Of Football* Reed International Books Limited 1996. p.11 ISBN 1-85613-341-9

[49] "Laws of the game (Law 7.3–The duration of the match)" (http://web.archive.org/web/20080603064822/http://www.fifa.com/flash/lotg/football/en/Laws7_03.htm). FIFA. Archived from the original (http://www.fifa.com/flash/lotg/football/en/Laws7_03.htm) on 3 June 2008. . Retrieved 3 March 2010.

[50] For example in the FA Cup prior to the semi-finals.

[51] Collett, Mike (2 July 2004). "Time running out for silver goal" (http://www.rediff.com/sports/2004/jul/02silver.htm). Reuters. Rediff.com. . Retrieved 7 October 2007.

[52] "Laws of the game (Law 15–The Throw-in)" (http://web.archive.org/web/20070913142556/http://fifa.com/flash/lotg/football/en/Laws15_01.htm). FIFA. Archived from the original (http://www.fifa.com/flash/lotg/football/en/Laws15_01.htm) on 13 September 2007. . Retrieved 14 October 2007.

[53] "Laws of the game (Law 16–The Goal Kick)" (http://web.archive.org/web/20070913141725/http://fifa.com/flash/lotg/football/en/Laws16_01.htm). FIFA. Archived from the original (http://www.fifa.com/flash/lotg/football/en/Laws16_01.htm) on 13 September 2007. . Retrieved 14 October 2007.

[54] "Laws of the game (Law 17–The Corner Kick)" (http://web.archive.org/web/20070913142324/http://fifa.com/flash/lotg/football/en/Laws17_01.htm). FIFA. Archived from the original (http://www.fifa.com/flash/lotg/football/en/Laws17_01.htm) on 13 September 2007. . Retrieved 14 October 2007.

[55] "Laws of the game (Law 13–Free Kicks)" (http://web.archive.org/web/20070913142645/http://fifa.com/flash/lotg/football/en/Laws13_01.htm). FIFA. Archived from the original (http://www.fifa.com/flash/lotg/football/en/Laws13_01.htm) on 13 September 2007. . Retrieved 14 October 2007.

[56] "Laws of the game (Law 14–The Penalty Kick)" (http://web.archive.org/web/20070913142717/http://fifa.com/flash/lotg/football/en/Laws14_01.htm). FIFA. Archived from the original (http://www.fifa.com/flash/lotg/football/en/Laws14_01.htm) on 13 September 2007. . Retrieved 14 October 2007.

[57] "Referee's signals: advantage" (http://news.bbc.co.uk/sport1/hi/football/rules_and_equipment/4188646.stm). *BBC Sport* (BBC). 14 September 2005. . Retrieved 4 March 2011.

[58] IFAB. "Law 5: Referee: Advantage" (http://www.fifa.com/mm/document/affederation/generic/81/42/36/lawsofthegame_2010_11_e.pdf) (PDF). *Laws of the Game 2010/2011*. FIFA. p. 66. . Retrieved 4 March 2011.

[59] "Confederations" (http://www.fifa.com/aboutfifa/organisation/confederations/). FIFA. . Retrieved 4 March 2011.

[60] The number of competing teams has varied over the history of the competition. The most recent changed was in 1998, from 24 to 32.

[61] "The FIFA Calendar" (http://www.fifa.com/aboutfifa/calendar/events.html). FIFA. . Retrieved 12 June 2010.

[62] "Football Equipment and History" (http://www.olympic.org/football-equipment-and-history?tab=1). International Olympic Committee (IOC). . Retrieved 4 March 2011.

[63] "Football – An Olympic Sport since 1900" (http://replay.waybackmachine.org/20090601015157/http://www.olympic.org/uk/sports/programme/index_uk.asp?SportCode=FB). International Olympic Committee (IOC). Archived from the original (http://www.olympic.org/uk/sports/programme/index_uk.asp?SportCode=FB) on 1 June 2009. . Retrieved 7 October 2007.

[64] "Event Guide – Football" (http://olympics.sportinglife.com/olympics/story/0,23911,14986,00.html). *sportinglife*. 365 media group. . Retrieved 5 March 2011.

[65] "Organising Committee strengthens FIFA Club World Cup format" (http://www.fifa.com/tournaments/archive/clubworldcup/japan2007/releases/newsid=570740.html). FIFA. 14 August 2007. . Retrieved 7 October 2007.

[66] Fort, Rodney (September 2000). *Scottish Journal of Political Economy*. **47**. pp. 431–455. doi:10.1111/1467-9485.00172.

[67] "Estudiantes win Argentina Apertura title" (http://msn.foxsports.com/foxsoccer/latinamerica/story/Estudiantes-win-Argentina-Apertura-title). *FoxSports*. Associated Press. 13 December 2010. . "Under the system used in Argentina and most of Latin America, two season titles are awarded each year – the Apertura and Clausura."

[68] Hughes, Ian (31 March 2008). "Premier League conquering Europe" (http://news.bbc.co.uk/sport2/hi/football/europe/7321408.stm). *BBC Sport* (BBC). . Retrieved 27 May 2008.

[69] Taylor, Louise (29 May 2008). "Leading clubs losing out as players and agents cash in" (http://www.guardian.co.uk/football/2008/may/29/premierleague). *The Guardian* (London). . Retrieved 28 November 2008.

[70] Gregory, Patricia (3 June 2005). "How women's football battled for survival" (http://news.bbc.co.uk/sport1/hi/football/women/4607171.stm). *BBC sport* (BBC). . Retrieved 19 February 2010.

[71] Alexander, Shelley (3 June 2005). "Trail-blazers who pioneered women's football" (http://news.bbc.co.uk/sport1/hi/football/women/4603149.stm). *BBC sport* (BBC). . Retrieved 19 February 2010.

[72] "Tournaments: Women's World Cup" (http://www.fifa.com/tournaments/archive/tournament=103/awards/index.html). FIFA. . Retrieved 11 March 2011.

External links

- Federation Internationale de Football Association (FIFA) (http://www.fifa.com/)
- The Current Laws of the Game (LOTG) (http://www.fifa.com/worldfootball/lawsofthegame.html)
- The Rec.Sport.Soccer Statistics Foundation (RSSSF) (http://www.rsssf.com/)
- Association football (http://www.dmoz.org/Sports/Soccer//) at the Open Directory Project

Article Sources and Contributors

Peter_Mollez *Source*: http://en.wikipedia.org/w/index.php?title=Peter_Mollez *Contributors*: Fram, Jevansen, Jogurney, Karma-AH, Koavf, Oqaz, Rjwilmsi, TFCforever, Waacstats, Welsh, Zombie433, 2 anonymous edits

R._Charleroi_S.C. *Source*: http://en.wikipedia.org/w/index.php?title=R._Charleroi_S.C. *Contributors*: 3zozey, Aecis, Afnal, Airconswitch, Am4387, Amirreza, Angelo.romano, BalkanFever, BanRay, Ben Ben, Beni93, BigDom, Bine Mai, Bobo192, Bracamonte, Bryan Derksen, Chanheigeorge, Clicketyclack, Colonies Chris, Conscious, DMG413, Danjsharp, Deadpunk, Defy, Djdannyp, Dribblingscribe, Eivind F Øyangen, Ev, Ez040, Galaxyharrylion, Geschichte, GiantSnowman, Gidonb, Gokhantig, GrahamHardy, Greenshed, Hairy Dude, HonorTheKing, Ilikeeatingwaffles, Im1a2, JT72, Jakas1, Jaysuc, Jevansen, Jimbo online, Jmorrison230582, Joao10Siamun, Jogurney, Julien Tuerlinckx, Jyusin, K4v3, Kafuffle, Karma-AH, Lightmouse, MSClaudiu, Madripper, Manneken Pis, MarkMysoe, Markh991, Masumisyan, Matthew hk, Monkee102, Monstradamus, NYC2TLV, Nodmaster, Palffy, Pelotas, Pennystar, Princeofpersia1, Prof bed, Qwer25, Raji1611, Retepretep, Rhand007, Roccm, RyanGerbil10, Saxbryn, Sir Sputnik, Siva1979, Soccerguinea, Tim!, Tiresais, Tobes00, TonyStarks, Truthanado, Ugen64, Willy turner, Zombie433, נחום127, 352 anonymous edits

K.S.C._Lokeren_Oost-Vlaanderen *Source*: http://en.wikipedia.org/w/index.php?title=K.S.C._Lokeren_Oost-Vlaanderen *Contributors*: Adik7swiony, Aecis, Ajdebre, ArtVandelay13, Avala, BanRay, Bine Mai, Bobo192, Bocanegra, Bracamonte, Brian Brandell, Buster7, Caledonian Place, Chanheigeorge, Chelseabob, CrnaGora, Cukiger, Dale Arnett, DenCA, Dj nix, Dribblingscribe, Eivind F Øyangen, Ez040, Finniemc, Firas1977, FvS, Geregen2, Geschichte, GiantSnowman, GrahamHardy, Gusulfurka, Hairy Dude, Ief, Im1a2, JT72, Jaysuc, Joao10Siamun, Johan Elisson, Jonathan32NL, Julien Tuerlinckx, Karma-AH, Kartmen, Kolins, Kweniston, LimoWreck, MM1912, MTLskyline, Maccabi Netanya Rules, Maccabit, Manneken Pis, Matanel ly, Matthew hk, Monkee102, Namiba, PAPYAN1992, Palffy, Peecee1978, Pelotas, Player bio, Prof bed, Qwer25, Raji1611, Raymond Cruise, Retepretep, Roccm, Rsc-Arni, RyanGerbil10, Saxbryn, Showie2002, SimonVerm, Siva1979, Tbhotch, The Thing That Should Not Be, Thomaslaza13, Tim!, Wizardman, Woohookitty, Ypetrachenko, Zombie433, 171 anonymous edits

S.V._Zulte_Waregem *Source*: http://en.wikipedia.org/w/index.php?title=S.V._Zulte_Waregem *Contributors*: A18919, Add92, Aecis, AndTheCrowdGoesWild, Andres, Andrwsc, Avni.yotam, BanRay, Benedictwest, Bine Mai, Blgeoverlord, Bobblehead, Bobo192, Bracamonte, Britwell, Caledonian Place, Chanheigeorge, Chelseabob, Cloudz679, Daddy Kindsoul, Doberman Pharaoh, ElommolE, Flewis, Footballanalizer, Freddy Guarin, GGib, Geregen2, GiantSnowman, GrahamHardy, Groovenstein, Grunners, Hairy Dude, Ilikeeatingwaffles, Im1a2, Jaganjac, Jmichaelgiles, Julien Tuerlinckx, Junichi, Karma-AH, Latouffedisco, Le Fou, Lightmouse, LimoWreck, MSClaudiu, MTC, Maklion19, Manneken Pis, Matthew hk, Mishoo89, Ms2ger, Noel baran, Opaq23, Papado2, Pelotas, Player bio, Qwer25, Retepretep, Ronaldvd, RyanGerbil10, Sir-Nobby, Siva1979, Skier Dude, Skitzouk, Smirnoff103, Special-T, Stealth107, Taarten, Tdudkowski, Themelle444, Tim!, Tocino, Wester, WilliamF1two, Wout81, Yavni, Zombie433, 163 anonymous edits

Arsenal_F.C. *Source*: http://en.wikipedia.org/w/index.php?title=Arsenal_F.C. *Contributors*: 03tgordo, 04arnols, 24framespersecond, 28steve, 2mr41h, 5 albert square, 70mD, 729, 96simccartan, A bit iffy, A p3rson, A3RO, ARSNL, Aa4, Aac74, Aaron Schulz, Ab762, Ab9db, Abdiali15, Abdullahbs, Abeg92, Abreuzinho, Academic Challenger, Acetic Acid, Aceviper110, Achangeisasgoodasa, Aciyokrocky, AdamD123, AdamJacobMuller, Adashiel, Aderi, Adidasnike619, Adrian Sampson, Adthebad12, Adzeds, Aecis, Aediasse, AfC, Afc4life2003, Afrodiziak, AhMeD BoSS, Ahoerstemeier, Aid1223, Akamad, Akinsope, Akmilo 21, Akumatatsu61, Al3xDeViL, Alasdair, AlasdairShaw, Albatross2147, Albeano2361, Alborzb, Alechuncho, Alex "bwfc" Latham, Alex "happy" Latham, Alex Selby, AlexWilkes, Alexandros, Alexcavell, Alexkid9, Alexpeckover, Alexwood3711, Alexx04, Ali K, Alias Flood, Alii h, Alishaladi, Allyrosa, Alnchn, Alozawi11, Alphabet14, Alphachimp, Alphalphanator1, Altzinn, Alvshill, Amalthea, Amir beckham, Amitch, An Siarach, An4rew, AndTheCrowdGoesWild, AndrePeltier, Andrea 93, Andredabassman, Andresp91, Andrew Levine, AndrewFyffe, Andrij Kursetsky, Andriyrussu, Andy Ramsay, Andy&jord22, Andycjp, Andypandy.UK, Andysteve, AngelOfSadness, Angela, Angmering, Angus Lepper, Aniten21, Antandrus, AntonioBecerraPiano, Antony1821, Anwarps, Apotsios, Ar558, Arangan, Aras-real, Arbero, Archanamiya, Arinsau, Arjenrobben16, Arkady98, Arnemann, Arrowny, ArsenaI, Arsenal, Arsenal mjh, Arsenal00, Arsenal02071, Arsenal14, Arsenal1802, Arsenal22, Arsenal414, Arsenal44, Arsenal4Europe, Arsenal666, Arsenal8039, ArsenalWiki, Arsenaldc1988, Arsenalfan18, Arsenalkid700, Arshavin the Gunner, Arsonal, ArtVandelay13, Artathegreat, Arteyu, Arthur Smart, Arthurrrrr12, Artichoker, Arusnak, Arwel Parry, Asad112, Asarlaí, Ashleygrint, Ashton Gate, Asmah01, Asserghozlan, Aude, Auntof6, Avantsweater, Avb, Average Earthman, Avicennasis, Avni.yotam, Avono, Aztec06, BRACK66, Bact, BadWolf42, Bagpuss, Bainyrues, BaldBoris, Ballon d'or, BanRay, Banana Fingers, Bandana6, Banes, Baney60, Baqu11, Barciur, Baristarim, Baros1, BarretB, Bastin, Batsi, Bazov, Bbatsell, Bcnviajero, Beamathan, Beansformickey, Beckford14, Beckham162, Becksguy, Beesforan, Ben Ben, Ben Crouch, Ben123123123, Bender235, Benduke16, Bento00, Benzmit, Berbatov345, Berryedd8, Bertan1311, Bevo74, Bib219, BigDom, BigHairRef, Bigbluefish, Bigemore, Biggieyankfan, Bigturnip, BillyBatty, Birelandb, Bjarki S, Bkwillwm, Blocci, Bloom6132, Blowerb2214, Bluecity, Bluedevil04, Bluelion, Blufive, Boardnim, Bob Castle, Bob Palin, Bobblewik, BobbyAFC, Bobo192, Bobo55, Bocanegra, Boing! said Zebedee, Boko 29, Boothy443, Bopalula, Born2edit, Bornintheguz, Bort08, Bosnianboy, Boucsetan, Bozman7000, Bracamonte, Bradstonpickle, Brian0918, Brichcja, Brighterorange, Brobby93, Bruce1ee, Bruno Torres, Bruno-ban, Bssc81, Bubas, Bubbagump24, Bucs, Buickid, Burnsey10, Bydand, C i d, CDSANTACLARA, CMacMillan, Cabbagery, Cacau Toledo, Cactus.man, CalJW, Callmecow, CambridgeBayWeather, Campl004486, Camw, Can't sleep, clown will eat me, Canderson7, CanisRufus, Canjbez, Cantthinkofagoodname, Captainbeecher, Carlwev, Casliber, Catisfit, Cdawwwg, Celtmist, Centrx, Cfailde, Cg29692, Chacha15, Chancemichaels, Chandan Guha, Charlie999, Chelseachav, Chemicalrubber, Chensiyuan, Chink182, ChipmunkYo, Chovain, ChrisTheDude, ChrisUK, Chrism, Christian75, Christopher Parham, Citz, Cjindustries, Cjmarsicano, Cjordan15, ClarkF1, Classick08, Clayjay, Clemondo, Clever Booy, Cliftonian, Cloudz679, Clout, Cmz007, Cochilles, Coconuts00, Cocytus, Codeine, CohenMD, Cometstyles, Commander Shepard, CommonsDelinker, Conversion script, Coolboy10127, Coolhawks88, Cordless Larry, Crazy Jebus, Crespo7, Criscione, CrokeyBoy, Crookesmoor, Crosscountrycpjon, Cs-wolves, Cswift2, Ctangeman, Cubsfan05, Curps, Cursive, Cyde, Cyril Washbrook, DEINNES, DGJM, DHooke1973, DIEXEL, DJ Bungi, DJ Clayworth, DJ Zoot, DMJohnston, DUCKISJAMMMY, DVdm, Da Jadmeister, Dabomb87, Daemonic Kangaroo, Dako, Dale Arnett, Daltanimal, Damaster-rapper, Dan.aliencheese, Dana boomer, Dancter, Daniel Case, Daniel arsenal, Daniel0212, Danielgreen1970, Danielt998, Danmantalis, Dantasu, Dapsv, Darius20, Darkson, Darth Nub, DarthInsinuate, Darvel4, Darwinek, Davehannabuss, Davetherave765, Davidthebeast, Davieja, Dbagdi92, Dbenbenn, Dcooper, Deathtopudding, Deb, Dekisugi, Deliboo24, Delivernews, Delusion23, Demiurge, Demonblade, Denisutku, Denkyu, Denniss, DethFromAbove, Devil melvin, Devilatheaven, Deville, Devoindahouse, Dffgd, Dgkdk, Dharmabum420, Diabytastic, Diadia Mancho, Didrik, Digilee, Digirami, Digitalme, Dimon449, Dineshcooper, Dingdong65, Dirk Valentine, Discospinster, Dispenser, Djdan90, Djdannyp, Djh1102, Djln, Djwest 047, Dlohcierekim, DlxEU, Dnaile2, Dokuzovski1, Dorkules, Dpmuk, Dr Girard, Dr Sachs, Dr. Blofeld, Dr31, DrKiernan, Draganjovicic, Dragases, Draktorn, Dreaded Walrus, Drmies, Dryazan, Dsapery, Dudesleeper, Dunnsta, Duttlibra85, Dweller, Dysprosia, E Pluribus Anthony redux, ESkog, Eamon06, Eazysteve, Ed g2s, Eddie6705, Eddieonline, Edeltoro, Editore99, Edroeh, Edward, Edwy, Edyuen888, Eeeby, Efeboy, Eivind F Øyangen, Ekingunel, ElPipita, Eldragon21, Eldumpo, Eliran Levi, Elly7, Ellywa, Eloquence, Elzar, Eminabe53, Enoch Wong, Enterandypandy, Epbr123, Epolk, Ericoides, Erminameraz, Erzengel, Esa91, Escape Orbit, Esk, Etchy93, Eubulides, Everard Proudfoot, Everyking, Excirial, Exir Kamalabadi, Extraordinary Machine, Extremer88, Exxolon, FF2010, Facudis, Fadzrulhisyam, Falcon8765, Farhikht, Fastily, Fatty01, Fazmah, FeanorStar7, FelipeTM, Ferrari2503, Fezpp, Fifa8181, Fifty Percent Normal, Fijipeace, Fil usa, Finns, Firien, Fish and karate, Flameviper, Flamurai, Flipbatijib, Flrn, Flynnbc12, Fmorton, Footballgenius, Footballyears, FootieAnorak, Footy-Fans, Footyfan99, Forbsey, Fountainj, Fram, Frankby, Freakofnurture, Fredbambridge, Fredd the Redd, Fredde 99, Freddyballo, Fskis, Fudoreaper, Fullysick bro 101!, Fuzheado, Fxfxfx x, Fys, G. Capo, G000se, GHe, GL3N, GSR05, Gabriele Deulofleu, Gagandeep, Gail, Gaius Cornelius, Galatian, Ganryuu, Gareth Owen, Gavincmbll, Gavindow, Gazzaboi2k6, Gcsewowbar, Gdo01, Geeman, Geneb1955, Geni, Geoking66, Georgeorwell, Geovanni10, Ghingo, Ghinma, GiantSnowman, Giants27, Gilliam, Gilo1969, Gimboid13, Gittens77, Gjt6, GoPurpleNGold24, Gogo Dodo, Gokhantig, Golbez, GoldDragon, Gooner14, Gooner1990, Gooneradam, Goonerak, Goonerth, Goonerz, Gopslayer74, Gpliotis, GraemeL, Graingy, Greenandbluesleeves, Greenshed, Gregz, Grethas, Griffinofwales, Grimlock81, Grover cleveland, Grutness, Gspurs, Gtdp, Gump Stump, GunnerKris, Gunnerkid, Gunners15, Guoguo12, Gurch, Gurt Posh, Gwernol, Gzgianfreda, Hadal, Haeleth, Haldraper, HalfShadow, Halibutt, Halmstad, Hammersfan, Hampsterman, Hannodv, Harabanar, Harald the Great, Harbee, Hariharan91, Haroon-kherr, Harry the Dirty Dog, Harrywalker, Hasanchoudhury97, Hashim-afc, Hashmander, Hassocks5489, Haukurth, Haza-w, Heavy1234, Heenan73, Hell9, Henry wang, Heracles2008, Hermiod, Hestemand, Hig Hertenfleurst, Hildanknight, Hillman15, Hintha, Hmains, Hoiwa852, Hokuto lover, Honbicot, Honette, HonorTheKing, HornetMike, Hornymanatee, Hotingzilla, How0014, Howcheng, Hugh McCloud, HughBryning, Husond, Hvn0413, Hydeblake, Hydrolix, Hytarr, I didn't push her, IJA, INkubusse, Ialsoagree, Ian1981, Ianeiloart, Ibnshaadi, Icemanofbarcelona101, Ichiuk, Ief, Iffty, Igomaa, Ilikeeatingwaffles, Illuminattile, Ilva, Imlikeaboss, Indon, Infernix, Inventm, InvisibleK, Iridescent, IronGargoyle, Irpen, Isak11, Isavevski, Isteadsimon, Italia2006, Iugin, Ixfd64, J. 'mach' wust, JB82, JDamanWP, JHK, JHMM13, JNW, JRRobinson, JStewart, Jaaahn, Jacek Kendysz, Jack10, Jackodamisno, Jacoplane, Jaganjac, Jagger99, Jairodz, JakeH07, JakeHT, Jakestedman, Jakiša Tomić, James Kemp, JamesTheNumberless, Jamescockerton, Jamespurs, Jamiebijania, Jarashi, Jarmez, Jasca Ducato, Jasmeet 181, Jasynnash2, Jaxl, Jayjaybee86, Jayk47, Jayson4all, JaysonG, Jazbag, Jbarta, Jcw69, Jdforrester, Jeandré du Toit, Jearl30, Jed keenan, JediKnut, Jeff G., JeremyA, Jeronimo, Jess Cully, Jevansen, Jh1442, Jhlee97, Jim Douglas, JimClark, Jimbo online, Jimbob615, Jimfbleak, JimmyMac82, Jkhtong1, Jklin, Jl85, Jmolloy4, Jni, JoanneB, Joe Bamber, Joe Decker, JoeA2580, JoeBlogsDord, JoeLatics, JoeSmack, Johan Elisson, John Carroll, John Quincy Adding Machine, John howard is very cool, JohnWittle, Johnchiu, Johndman, Johnelwaq, Johngooner, Johnleemk, JohnnyPolo24, Jojhutton, Joker8910, Jonathanfear, Jonzi9, Jordanson, Jorsha, JoshDuffMan, Josquius, Jossi, Joy, Jpboyrox, Jrincewind, Jrockley, Jscales06, JtotheROC, Juki8007, Juliancolton, Jvallerine, Jw2034, Jwissick, K-KYL3-K, K-UNIT, K33wud, KBONETOPMAN, Kabeer rock, Kaizero, Kalaha, Kalleponken88, Kanabekobaton, Karafias, Karenjc, Kaypoh, Kbthompson, Kcordina, Keenos, Keith D, KeithD, Kelly Martin, Kengi1982, Ketiltrout, Kevin B12, Kevin M Marshall, Kevin McE, Kevinbi2004, Kevy2004, Kezzerkid, Khalid red, Khanas, Khaosworks, Khorshedalam, Khukri, Kiagla, Kidbrother, Kidstylez, Kieran96336, Kiero15, Kietotheworld, Killerkweller, King Dracula, King Gunner, King of the North East, Kingbrian123456789, Kingdriscoll, Kingfisherswift, Kinggene14, Kingjamie, Kingjeff, Kisuga, Klasnicinhk, Klenky, Klinean, KnowledgeOfSelf, Knucmo2, Koavf, Koeilouis, Kolins, Koperino, Kopfballungeheuer, Kotjze, KrakatoaKatie, Kronicz, Krraj, Krun, Kuemmjen, Kumordzi, Kungming2, Kurtziya, Kuru, Kushal2091, Kwamikagami, Kwig92, KyleRGiggs, Kynen, LC1250, LCmucky, LT78666, Lalaman, Lapafrax, Largom4, Latics, Lcb91, LeaveSleaves, Lectonar, Ledge.87, Lee Gregz, Legend27, Legendkillerpunk, Lemonade51, Lestermay, Leszek Jańczuk, Levi Versus, Lewisvenes, Lewit, Lex3000, Lightdarkness, Lightmouse, Ligulem, Lihaas, Lilkimloving, Linkinstreet, Lipik, Lironos, Little-big1, LittleOldMe, Liverpool Scouse, Liverpoolrulez, Lloydf640, Lmcintyre1, Localzuk, Longnecklongbones, Lordkazan, Lorenzarius, Loud Mouth Soup, Loustar99, LowVelocity, Lpeters001, Lucha-Method, Luk, Luke big, Luke570, Lukerpool, Lumpy3, Luna Santin, Luokehao, Lupin, Lyswim, MA5, MC MasterChef, MDCollins, MER-C, MK8, MRSC, MSJapan, MYS77, MZMcBride, Maad Dogg 97, Maartenvdbent, Maccabit, Maciekced, Madridista23, Madshurtie, Madw, Magister Mathematicae, Maias, Mailer diablo, Mairi, Maja123456, Malcolmarsenal, Malo, Malpass93, Malvolio.Blanco, Manishkeshwani, Mansmithyutd, Manu4eva, Manuiyer, MarSch, Marianocecowski, Mark272, MarkGallagher, MarkMysoe, MarkS, Markbrus, Markie111, Marlarkey, Martarius, Martin tamb, Martin707, Master Jay, MasterHasan, Masterjamie, Masuk50, Matt Adore, Matt Yeager, Matthew hk, Matthewairs, Matthewb86, Mattmanorient, Mattrhames, Mattsmithgb, Mattythewhite, Mav, Mav2001, MaxSem, Mayur, Mazito, Mcaparrotta, Mccabem32, Meey11, MegX, MeltBanana, Meme feme, Merovingian,

Messiarg10, Metagraph, Methnor, Metroponce, Mezza16, Mgdavison, Mgreen09, MichaelStorey, Michaelscullion1, MicroX, Midgrid, Mike Christie, MikeLebo, MikeW1989, Mikemor92, Mil Falcon, Millahnna, Millbanks, Millertime, Milloy, Minkhairov Arthur, Minoru, Mintguy, Mirage5000, MisfitToys, Miss Madeline, Misza13, Mithridates, Mjefm, Mltinus, Mo0, Modulatum, Molerat, Moonschool, Moontvrules, Moonty, Moreschi, Morhveem, Mortene, Mosmof, Mozzerini, Mpassman, Mpj17, Mr Cool, Mr Hall of England, Mr Stephen, Mr Window, Mr mormon, MrH, MrMagoo, Mshalfon, Mswake, Mufka, Mugglekiller, Mummy64, Muppet, Murp64, Musicandnintendo, Mustafur, My-dog-is-shep, Mymap, Myrmecophagous, Mórtas is Dóchas, NSLE, Nabs, Nakon, Narcis90, Narcisse, Nattywatty, NawlinWiki, Ned786, Nedward, Neil Auckland 10, Nemzneezy, NeonMerlin, Netoholic, Nfitz, Nh4444, Nhan nguyen tran, Niall123, Nick, Nick braz, Nickhooker1, Nicklansley, Nickolas.butti, Nickyp uk, NihalKoshy, Nika mardaleishvili, Nimicitor, Nish81, Nix D, Njh@bandsman.co.uk, Njm05, Nmarra01, No 7 man, NoIdeaNick, Nomist, Norbert2000, Norm mit, Northumbrian, Notzaj, Nowyouseeme, Nstrayert, NuclearWarfare, NufNufNicky, Nufc2006, Nuggets, Nuggets97, Number 57, O'Dowell, ORLANDORICHARDSON, Oabel, Obarskyr, Ocrasaroon, Odie5533, Ohconfucius, Oldelpaso, Ollywally, Olorin28, Omar 89, Omicronpersei8, Omphaloscope, Or12, Orangina2, Orion6767, Orphan Wiki, Osbus, Oscar86, Osvaldo88, Otaku-jin, Ou tis, Ouishoebean, Oun, OwenBlacker, Oztaz, P Ingerson, P0rk0-2diie4, PET, PRDH, PS., PS12, Pah2you, Pais, PakistanGangster, Pal, Papongza, Parkinson24, Pasado manana tu madre, Pasini, Pathmicphapar, Paul W, Pauldewart, Paulmdavies, Paulofthebakers, Pawelbal30, Paynemi, Pcpcpc, Peanut4, PeeJay2K3, Penguinsix, Pereirab04, PersianScarface, Peruvianllama, Peter Chastain, Peterinns, Peterman, Petree13, Petsky, Pgk, Phantom Thief, Phil Boswell, Phillystile, Phinnaeus, Pigsonthewing, Pilotguy, Piyush05, Pjharding24, Pkchan, Plasticup, Plastikspork, PlatinumJester, PlayCollective, Pmberry, Pobink, Pointer1, Poker Flunky, Politepunk, Poly87, Poppy, Portillo, Postlebury, Poulsen, Powelldinho, Powerhitter04, Ppntori, Prin, Prince Godfather, Pro66, Profoss, PseudoSudo, Psveindhoven, Pullshapes, Pwimageglow, PythonX, Pzpz, QmunkE, Quadrius, Quentin X, Quikfit, Quincy042, Qureshiao, Qwertyuiop1262, Qwghlm, R Lowry, R'n'B, RCS, RJC, ROMAM20, RRR22, RaNdOm26, Rahman1994, Rahman9420, Raiokuda, Ral315, Rambo's Revenge, Rancit, Randomgeezer, Rares992, RateTheRef, Raul654, Raven4x4x, Ray30, Razororz, Rbrwr, Rdd446, Reach Out to the Truth, Red&White4life, RedmanJB, RednessInside, Redwolf24, Reedy172, Refsworldlee, Rehcuobolim, Reindeer, Rellis1067, Renesis, RescueRanger702, Retepretep, Rettetast, RexNL, Riahc3, Ric36, Rich Farmbrough, Rich nffc, Richard Lee 9, RichardBagnall, Richlaughlin, Ricky Sen, Rigel1, Ringaw7, RissederEngel, Rjcb23, Rje, Rjwilmsi, RobertG, Roberto Flamingo, Robriotuk, Robsinden, Robwingfield, RockMFR, Roleplayer, Roosterrulez, Roryjburke, Ross1, Rosskey711, Roundface, RoverTheBendInSussex, Royalguard11, RubberJoshy, Rupertslander, Rutld001, Rviktorb, Rvp11, RxS, Ryan Taylor, Ryandell12345, Rye1967, Ryecatcher773, S M T1988, S980182, SE13Gooner, SE7, SEWilco, SMC, SQGibbon, ST47, Saebhiar, Salomonis, Salyale, Sam Korn, Sam wade, Samir Nazari, Samw91, San1140, Sanchez196, SandyGeorgia, Santaclaus21, Santryl, Sars, Sarumio, Sashe, Satsumas, Scaife, Scatesy, Sceptre, Schalkcity, SchuminWeb, ScottDavis, Scottb999, Scproductions, Screech123, Sdprodigy, SeanMack, Sebastiankessel, Secretlondon, Selmo, SenorKristobbal, Sephjnr, Sevenplusone, Sh17break, Shadowjams, Shahin.shn, Shaidar cuebiyar, Shakehandsman, Shamboko, Shanel, Shanes, Shannonnights, Shaq555, Shenandora, Shereareraer, Shikinluv, Shitehead, Shobs10, Shooter16101, Shortride, Shpanda, Shreyasm89, Shu1234, Shylocksboy, SiAFC1, SilkTork, Silly racoon ugh, Sillyfolkboy, SimonMayer, Siqbal, Sir Nicholas de Mimsy-Porpington, Sir Stanley, Sirchrisjbrown, Siva1979, Sjakkalle, Sjc, Sjjames, Skitzouk, Skooing, Skyblueshaun, Slevenback, Slnaing07, Slow Graffiti, Slowking Man, Slumgum, Smacey, Smahoney1981, Smon, Snake eyez89, Snaveuk, SndrAndrss, Snowolf, Snoyes, Snypa89, Soccerpunkrocker, Someoneinmyheadbutit'snotme, Sparrowgoose, Spartan008, Spasomabaso, Spellcast, Sphinx06, Spiritofsussex, Sportyman1970, Squares18, Srikeit, StAnselm, Staffa93, Star rocker, Starnestommy, Steel, Stephen Bain, Stephen MUFC, Stephenb, Steve newbold, Steve tutty, SteveO, Steveweiser, Stevodonnelly, Storageking, Strawberry on Vanilla, Stripyfab, Strop, Struway2, Stubblyhead, Suckermonster, Sulo100, Sunil060902, Sunray, SuperHamster, Supergunner08, Supreme nutter, Supsap, Svick, Swanbiff, Swatjester, SxB, Syahmi123, Syjytg, Symo-EFC, Szalas, T-jay101, TBadger, TKK, Tabletop, TaerkastUA, Tagishsimon, TakTak, Takki the Greek, Talrinys, Tangerines, Tangotango, Tanweijin, Taroaldo, Tawker, Tbhotch, Techsplash, Tedaram, Teddypickle, Template namespace initialisation script, Terence, TerriersFan, Tfe, The C of E, The Great Arsenal Fan, The Land, The Rambling Man, The Thing That Should Not Be, The sunder king, The1yathindran, TheBigJagielka, TheDoctor10, TheKMan, TheMongoose, Theallian, Theda, Theenglishway88, Theiceman10, Themza, Theodore144, Theodorevp, Theone010203, Thepbc, Thepenpen, Thepredatorhandshake, Therefore, Theresa knott, Thesilentassassin, Thesmurf85, Thewayforward, Theweapons, Thierry474, Thierryhenry1000, Thisiskieran, Tholynski, Thomasmc14, Thorpe, Throup, Thumperward, Thunderbrand, Thunderstats, Tiago Heitor, Tide rolls, TigerShark, Tim!, Tim8008, Timberframe, TimmyFc, Tinygunnerlee, Tip123, Titoxd, Tkko, Tmacgoill, Tmol42, Tocino, TodorBozhinov, Toge1988, Token 000, Tom harrison, TomaTomaDJ, Tomd89, Tommy2010, Tomo7797, Tomsega, Tomtom1156, Tony1, Tonynz, Totalsoccer, Tottingham, Tpth, Travelbird, TreasuryTag, Treegoatman, Trekkie459, Trelawnie, Trident13, Tripod86, Troels Nybo, Tropicalresources, Truegooner28, Trumpkin, Truthofangels, Tubz27, Turkish, Turkish van, Tuxnduke, Twiceuponatime, Tyndie, USASkiWolf, UberScienceNerd, Ugen64, Ugur Basak, Ultracanalla, Ultras GE, Ultratomio, UncleTupelo1, Uncleshola, UnicornTapestry, Unitednut, Universal Hero, Unreal7, Uskboy, V louis, VEO15, Vald, Vanderdecken, Vanig14, Vanished895703, Vardion, Variance, Vaubin, Veila, Veldorome, Velociraptor888, Vendettos, Ventusa, ViTo13, Vic1234567890, Vickerman625, Victor Zheng, Vikrantpal, Villa88, Vincej, Violetriga, Vipinhari, Viridae, Viridian, Vishmaster, Vivbaker, VodkaJazz, Vrenator, Vssun, WATP, WILLKW, WMFLIX, WPArgentarius, WR227, Walter Görlitz, Wansell, Wardy13, WarthogDemon, Wayne Slam, Wayward, Wdyoung, Webb1175, Wee Man, Wfgiuliano, Wicka wicka, Wiggy04, Wiki alf, WikiWitch, Wikieditor06, WikipedianDog, Wikipete, Wikiplayer13, Will Beback, Will Beback Auto, Will Colwell, WillMak050389, William Pietri, WilliamF1, Willrey619, Winchelsea, Wislabe, Wknight94, Wokky, Wolf1, Wonderrrrrrrrr, Wonderwizard, Woody, Woolwich, WorldsFastestPakeha, Woseph, Wrestler9000, Wswaine, X96lee15, XTomScottx, Xelo16, Xezbeth, Xftr7700, Xhienne, Xiner, Xobxela, Yafar, Yanksox, Yashiv, Yavni, Yaya Silva, Yelgrun, Yellow up, YellowMonkey, Yeunk, Yohan euan o4, Yonatan, Young gunz10, Yousou, Ytfc23, Yunghei, Zack, Zanoni, Zbase4, Zeitgeist, Zero1328, Zip600001, Zombie433, ZooPro, Zooopzooop, Zorcho, Zsinj, Zzizoo, Zzuuzz, Zzyzx11, Zzzzz, ^demon, Ævar Arnfjörð Bjarmason, ΟΣΦΠ, Александър, Дикий Ёжик, ילוקורב, 2616 anonymous edits

Club_Brugge_K.V. *Source*: http://en.wikipedia.org/w/index.php?title=Club_Brugge_K.V. *Contributors*: A18919, Aecis, Alan McBrazil Burger, Alarix, Andrwsc, Angelo.romano, ArtVandelay13, Arwel Parry, Avala, Ayomaju, Azon789, Balerion, Bartekos, Bastin, Bencherlite, Bertzor, Bine Mai, Blakewiki, Bobaggins, Bobo192, Bort08, Brugselaar, Bruno-ban, Bubas, Chanheigeorge, CipherPixel, ClubBrugge, Colonies Chris, CommonsDelinker, Crown1776, Cs-wolves, Cursive, Curswine, Cybermb, Dale Arnett, Daniel sf, DanielCD, Dapsv, Darius Dhlomo, Dbenbenn, Deadpunk, Deki147gta, Dewritech, Dimon229, Dirkinho, DrFishcake, Drilnoth, Dryazan, Ed g2s, Euku, Everton, Ez040, Fetofs, Fram, Frietjes, G000se, Galaxyharrylion, Geregen2, Geschichte, GiantSnowman, Grant.Alpaugh, Hairy Dude, Harbee, Hashishu boy, HonorTheKing, Hovhannesk, IceKarma, Ief, Ilikeeatingwaffles, J.delanoy, JLaTondre, Jaellee, Jed, Jogurney, Johan Elisson, Julien Tuerlinckx, Kalaha, Karma-AH, Kerbaby, Kingjeff, Kostisl, Kumordzi, Latouffedisco, Le Fou, Levon1992, Lightmouse, LilHelpa, LimoWreck, LyXX, MSClaudiu, Manneken Pis, Materialscientist, Matthew hk, Mattythewhite, Mayumashu, Mentoz86, MiLo28, Michfan2123, Monkee102, Motor, Mr Hall of England, Ms2ger, NFLNBA, Noel baran, Oliviu, Pal, Palffy, Pelotas, Pino2804, Poulsen, ProudTarjaholic, Red Winged Duck, Retepretep, Roeeyaron, Rosenborg, RoverTheBendInSussex, RyanGerbil10, Saint9016, Salgueiro, Salters, Screen five, Selket, Sharontampon, Sherool, Sir Sputnik, Siva1979, Skyblueshaun, Stekke01, Stephan Sensuality, Stitchface, Struway2, Syjytg, TPDossche, Taarten, Tfcfan10, Thecrookedcap, Tim!, Timbouctou, Tocino, Tody copp, TonyStarks, Ultras GE, VEO15, Wangi, Wereon, Wes Duschek, West Brom 4ever, Whkoh, WikHead, Wislabe, Woohookitty, Wuhwuzdat, Xenomorph1984, Ykpdr, ZloK, Zombie433, 387 anonymous edits

K.V._Kortrijk *Source*: http://en.wikipedia.org/w/index.php?title=K.V._Kortrijk *Contributors*: Aecis, Asaikana, BanRay, Behrami, Bine Mai, Bobo192, Bracamonte, BrianBird, Calaka, Caledonian Place, Chelseabob, ChrisHodgesUK, Dirkinho, Eivind F Øyangen, Geregen2, GiantSnowman, Giovanny Saelens, GrahamHardy, Greenshed, Ief, Im1a2, Jaganjac, Johan Elisson, Julien Tuerlinckx, Karma-AH, Lamadude, Latouffedisco, LilHelpa, Lucha-Method, Ludde23, MSClaudiu, Madcynic, Manneken Pis, Masumisyan, Nzd, Pelotas, Prof bed, Rapsar, Raymond Cruise, Retepretep, Rsc-Arni, Secretaria, Shailevi8, Siva1979, Sparxx1234, Taarten, Thuglife99, Tim!, TonyStarks, Welsh, Wislabe, Wizardman, Zombie433, 127 anonymous edits

Football_in_Belgium *Source*: http://en.wikipedia.org/w/index.php?title=Football_in_Belgium *Contributors*: 16@r, Antiochos4, Arkangel lucifer, BertK, Buster7, Daemonic Kangaroo, Djln, Drpickem, DéRahier, EoGuy, Eugene-elgato, Everyking, Fram, GAB7012, GraafGeorge, Gregorybean, Hmains, JH-man, Jcarls1, Julien Tuerlinckx, JustJoe, Kanthoney, KathrynLybarger, Lamadude, LilHelpa, LimoWreck, Manneken Pis, Maxime.Debosschere, Mohrflies, Nach0king, PeeJay2K3, Pelotas, Philippe, RainbowOfLight, Rich Farmbrough, Stuartymack, WATP, Whistopathe, 75 anonymous edits

Association_football *Source*: http://en.wikipedia.org/w/index.php?title=Association_football *Contributors*: 'Arry Boy, 03BTHOMPSON, 05fcrane, 10metreh, 1337B3A57, 144.132.75.xxx, 194.109.232.xxx, 1burke, 1crywolf1, 200120037, 209.20.225.xxx, 21655, 24.218.142.xxx, 24fan24, 2writer, 4019jimmyw, 540dash, 62.253.64.xxx, 62.7.47.xxx, 62.7.5.xxx, 63.192.137.xxx, 84user, A-giau, A.a679100, AEMoreira042281, AJCham, AJS2050, ARrohetMeZemer, Aabha R, Aaron1716, Abc123youwantme, Aberdeen fc, Abigail-II, Abrech, Abu badali, Academic Challenger, Acalamari, Accurizer, Achinn, AdamWeeden, Adambro, Addshore, Adizlaja, Admrboltz, AdrianTM, Adzeds, Aeon1006, Aeons, Aesopos, Aftesk, Agentknight, AgnosticPreachersKid, Aheyfromhome, Ahills60, Ahoerstemeier, AileanMacRaith, Airconswitch, AirdishStraus, Aitias, Aj hicks, Ajaxkroon, Akamad, AkdenizliAslan, Aksi great, Al E., Aladdinlee, Albinomonkey, Ale jrb, AlefZet, Aleksandr Grigoryev, Alex.muller, Alex43223, Alexis, Alexisamattlover, Alexsh, Alexsoccer102, Alexwiki69, Algebra, Alias Flood, Alsandro, Amalthea, Amazonien, Ambrose222, Ambuj.Saxena, AmiDaniel, AmishThrasher, Amitch, Amorymeltzer, Ams80, AnOddName, AnakngAraw, Anand Bindra, Anaraug, AndeanThunder, Anders Torlind, Anders Törlind, Andocomando, Andonic, Andreasegde, Andres, Andrew Levine, Andrewlp1991, Andries, Andrij Kursetsky, Andrwsc, Andy Marchbanks, Angelo.romano, Angles theatre, Angmering, Angus Lepper, Angusmclellan, Anish9807, Anonywiki, Antandrus, Anthony, Anthonyd3ca, Anyabean17, Aplomado, Aranherunar, Arbor, AriGold, Arnemann, Art LaPella, Arthena, Arthur Holland, Arwel Parry, Asdackla, Ashesdhakal, AshleyMorton, Asmah01, AstroNomer, Astrovega, Atari 667, Athaenara, Atharshiraz, Athomas7990, Attilios, Aubtiger2008, Aude, Auntof6, Australian Matt, Autoerrant, AutomaticWriting, Autonova, Auxil, Avitaltr, Avs5221, Awesome511, AxelBoldt, Aymatth2, Az1568, Azzurro2882, B.Wind, B4hand, BACthegreat, BBJAPS3, BGTopDon, BKH2007, BRG, BUblogger, Babylon32, Baconfish, Badgernet, Bagpuss, Baldhur, Ballboys10, Balthazarduju, Balwan, Bananaman69, Bangaluru, Banhtrung1, Barcafan10, Baristarim, Barliner, BarretB, Bart133, Basawala, Basbll4lif23, Baseballbrett13, Bastin, Batman2005, BatteryIncluded, Bawolff, Bbatsell, Bearbug21, Beeman20, Behzad20, Ben davison, Ben-Zin, Ben9dan16, Bender235, Bennelliott, Bennybp, Bento00, Betelgeuse, Beve, Bevo, Bevo74, Bgates456, Bhrdkor123, Biatch, BierHerr, Big Jock Knew, Bigdumbredman2, Bigearedtiger, Bigeazy, Biggieyankfan, Bigpapawaka, Bikeable, BillWSmithJr, Billward, Bkitko, Bkonrad, Blarg000001, Bleakcomb, Bletch, Blokee, Blotsfan, Blue520, Bluelion, Boavi, Bob Palin, Bob schriver, Bob199393, Bob512012, Bobaldinho10, Bobblewik, Bobo192, Bobomg, Bobryuu, Bogdangiusca, Boing! said Zebedee, Bomchickawah, Bongwarrior, Boojanam01, Boole, Borisblue, Bornfury, Bornintheguz, Boshtang, Boychoir, Bozzzzz, Brad101, Branddobbe, Brandeis, Branko, BreenJnr2@aol.com, Breezeonhold, Brendanurbanwarrior, Brent william mccoy, Brentonf, Brian0918, Brickie, Brighterorange, Brion VIBBER, Brisngr, BritishWatcher, Brod7, Brossow, Bruce1ee, Bryan Derksen, Bryansprang, Bssc81, Btljs, Bubba hotep, Bubbla, Buchanan-Hermit, Bullzeye, Bunnyho1, Bwithh, Byrdman15, ByteofKnowledge, Bz2, C.Fred, C12H22O11, CJ, CPMcE, CS.Aussie, CWii, Cactus.man, Cafzal, Caharayo, CahirAndIrish, Calsicol, Caltas, CambridgeBayWeather, Cameron5dollars, Camerong, Can't sleep, clown will eat me, CanbekEsen, Canek, Canjth, Captain-n00dle, Card Football King, Cardsplayer4life, Carnun, Carreira, Casablanca2000in, Casey56, Casper2k3, Castlecraver, Cataphract, Catch47, CatsClaw, Cbrown1023, Cdc, Cedders, Centrx, Ceoil, Cfarzan, Cfcrowdies18, Cfrydj, Chabby, Chalkwalker, Chandler, Chanheigeorge, CharlieZeb, CharlotteWebb, ChaucerGeoffrey, Checkmate911, Chibinho, Chill doubt, Chillum, Chinesekage, Chingiss, Chkno, Chodorkovskiy, Chris Roy, ChrisTheDude, Chrishatch1973, Chrislk02, Chrism, Christpunkergirl, Chrmb, Chromancer, Chuq, Citizen Premier, Ckatz, Clamster5, Clawed, Clio64B, Clngre, Clocksarecool, Clout, Cls14, ClubOranje, Clydecoast, Cmdrjameson, Cncxbox, Cobaltbluetony, Cobaltcigs, Cocoaguy, Code E, Code phil, Codex Sinaiticus, Cold Season, Colin stuart, Colin9brown, Colipon, Collard,

Colonies Chris, Commander Shepard, CommonsDelinker, ComputerGuy, Conscious, Conversion script, Cookiva, Coolcaesar, Corean2006, Corvus cornix, Costlab, Counsell, Courtney03, Crafedog, Craig Staples, Craigman0101, Craitman17, Crazycomputers, Crazytales, CreerBFC, Cribananda, Crissov, Cristian Cappiello, Crosstimer, Crystallina, Cs-wolves, CtrlAltDek, Cuchullain, Cuecho, Curps, Custodio.oliveira, Cutler, Cuvette, Cyan, Cybercobra, Cybershore, Cyp, DANE RAMADAN YOUSSEF, DARTH SIDIOUS 2, DB, DGJM, DIEXEL, DJ Clayworth, DOwenWilliams, DVD R W, Da Mastaa, DaGizza, Dachamp724, Dacxjo, Daddy Kindsoul, Daduzi, Daemonic Kangaroo, Daftpunkboy93, Dahliarose, Dale Arnett, Damienhunter, Dan Atkinson, Dan the man 10, DanKeshet, DanRUK, DancingPenguin, Daniel, Daniel Case, Daniel.Cardenas, Daniel5127, Daniel987600, Danlina, Danltn, Danreitz, Dantedanger, Dantheman0056, Dantheman531, Danthemankhan, Daquios, Darkildor, Darkrifle, Darry2385, Darthgriz98, Dave.Dunford, Dave101, Daveb, Davewild, David Johnson, David Levy, David Schaich, Davidbrookesland, Dawnseeker2000, Dc19, Ddfree, Deadlock, Deadmanwalking1, Deanh, Debate987, Deeb, Delirium, Delldot, Deltabeignet, Den fjättrade ankan, Dendodge, Denhamd2, DerHexer, Derek Ross, Derild4921, Deus Ex, Dgw, Didier13, DiegoTristan, Differentgravy, Digitalme, Dijon312, Dim SIM.StPaTs, Dimitrakopulos, Dina, Dinis80654, Dinshoupang, Dirkbb, DirkvdM, Disavian, Discospinster, Disorganized 676, Diving2010, Djdan90, Djln, Djmadd93, Doc glasgow, DocWatson42, Docu, Document, Doglover77, Dojarca, Domcabcd, Donama, Donatj, Donfbreed, Dostal, Doubleslash, Downloadmeh, Dr Santa, Dr31, DrKiernan, Dradious, DragonHawk, DragonflySixtyseven, Dragonnas, Draicone, Drmies, Dru of Id, Dryazan, Dudesleeper, Duja, Durin, Dust Filter, Dustimagic, Dweller, Dycedarg, DylaHeary, DylanW, EBL, EEMIV, EJF, EPs, ERcheck, ESkog, EWS23, Eagle4000, Earl CG, East718, Ebeisher, Eceresa, Eclecticology, Ed Poor, Ed g2s, EdC, Eddyji999, Edgar181, Eielsonafbman, Einstein100, El Suizo, Elliskev, Emc2, Empty Buffer, Emura, Encephalon, Endroit, EndymiOn, Energyfreezer, Enigmaman, Enoch Wong, Enviroboy, Epsteada, Equazcion, Erath, Ergmuncer, Ericoides, Error, Erzengel, Esanchez7587, Esperant, Estel, Eu.stefan, Evekrogman, Evenmadderjon, Evercat, Everyking, Evice, Evillution, Ewen, Executive.koala, Exploding Boy, Ezeu, Ezhiki, FF2010, Fabian9, Facts707, Fadf, FaerieInGrey, Fainites, Falcon8765, Fan-1967, Fang Aili, Fanta206, Fastily, Fasttimes68, Father Goose, Favonian, FelisLeo, Felix Portier, Fernandopascullo, Feroang, Fewskulchor, Ffahm, Fieryspider, Finell, First Light, FisherQueen, Fleagle11, Fliffy, FlipCockle, Floodamanny, Floridan, Fluffernutter, FluffyWhiteCat, Fluke08, Fmhsxc2010, Fonsildiamond, Fonzy, Football Fan Zone, Footballhead, Footeditor, ForeverWhiteRose, Forshizzle, Foxlarocks, Fr. Jim, Fragbase, Fraggle81, FrancoGG, Frank87, FrankCostanza, Freakofnurture, Fredrik, Freepenguin, FreplySpang, Frischy, Frood, Froogoleer, Fruit666, Frymaster, Funandtrvl, Funknl, Funnyfumble5, Fuzzle, Fvasconcellos, Fvw, Fyyer, G-Man, GHoeberX, GVnayR, Gar, Gareth Owen, Garzo, Gaulwiki, Gavinho, Gaw jess, Gazcheung, Gciriani, Gdarin, Gdo01, Gene Nygaard, Geoff janes, Geopersona, George The Dragon, Georgia guy, Gergerballball, German Player, Gerrit, Ghosts&empties, GianniG46, GiantSnowman, Giants2008, Giff1002, Gilesz182, Gilliam, Gilmore Guy, Gilo1969, Gimboid13, Ginajanota, Girlslutgirl, Gizmoleeds, Gkhoyt, Glass Sword, Glen, Globalsolidarity, Gman124, Gnangarra, Gnevin, Goatherd, Godneck, Gogo Dodo, Gohoos13, Gohoos131, Goldfinger820, Gollumz001, Gooface, Gordongeiger, Grabbies200, Grace Note, GraemeL, GraemeLeggett, Grafen, Graham87, Grant.Alpaugh, Grant65, Green, Greenman, Gregoryj83, Grey Shadow, Gridlock Joe, Griffinofwales, Grm wnr, Grondemar, GroveWanderer, Grover cleveland, Grunners, Gun Powder Ma, Gunray, Gurch, Gwcstcs, Gwernol, Gytrem, Gz33, Hadal, Hadger, Hagerman, Hairchrm, Hairouna, Hairy Dude, Hajima, Haldraper, HalfShadow, Halogeek, HandsomeFella, Happy2, HappyCamper, Happysailor, Harry Hotspur, Harry Potter, Harry19023, Hashimjaved, Haukurth, Hawkeye7, Hayden5650, Hegemon7786, Heightwatcher, Hemanshu, Henrygb, Hephaestos, Herbert Shek, Herbertxu, Heron, Hesamj, Hesperian, Hflegit16, Hgrenbor, Hi little kids, Hig Hertenfleurst, HighKing, His Ryanness, Hmains, Hockeyplayer371, Hollandfan, Holyjoe, Homer slips., Homestarmy, HomieG2008, Honette, HonorTheKing, Horologium, Htews, HungrySphynx2241, Husky, Husond, I already forgot, IJK Principle, INkubusse, Iahead, Iamajpeg, Iamthecutestboyever, Iangibbins, Ichinisan Ilu, Igiffin, Igor, Ihatenyy13, ImperatorExercitus, Imroy, Inhumandecency, Innotata, Intelati, Intellectualfive, Inter16, Interchange88, Iohjasdb, Ionutzmovie, Iota, Iran11, IronGargoyle, Isadera, Isfalk, Itsmine, IvoJovo, Ixfd64, Izelpii, Izno, J Di, J Hofmann Kemp, J.delanoy, JCam, JDDJS, JForget, JFreeman, JHMM13, JIP, JJGD, JLINCON, JNW, JRHorse, JSpudeman, JStewart, JYolkowski, Jacek Kendysz, Jack forbes (renamed), JackieMa, Jackosta1234, Jacoplane, Jacquerie27, Jaffer, Jahiegel, Jake2392, James Foulds, James530, JamesBWatson, JamesMLane, Jamesr@thesportbar.com, Jandrews23jandrews23, Jaranda, Jason Potter, Java13690, Jaxl, Jchild, Jcrow43, Jdawg4820, Jed, Jeff.nolan, Jeff3000, Jennie--x, Jenny4life, JeongHW, JeremyA, Jeronimo, Jersyko, Jesismael, Jesus182, Jfg284, Jhenderson777, Jibbajabba, Jiggelmaster7, Jimfbleak, Jimgawn, Jklin, Jmcc150, Jmed123, Jnestorius, Jni, JoanneB, Joe Cannon Fan, JoeLatics, Joebamber1990, Joebloggsy, Johan Dahlin, Johan Elisson, John, John Anderson, John Reid, JohnOwens, Johnmc, Johnny Jane, Johnnyonthespot, JonBroxton, JonC, JonHarder, Jonik, Jonnysamleland, Jooler, Jor70, Jose77, Joseph Solis in Australia, Joseph Wajsberg, Josh167, Josh619112345267, JoshDuffMan, Josquius, Jossi, Journalist, Joy, Joyous!, Jpogi, Jpollar, Jrdragster11, Jrodes, Jtdirl, Jtkiefer, Jtle515, Ju66l3r, Juancamaney22, Jujutacular, Julia Brewer, Julianp, JuneGloom07, Jusdafax, Jwissick, KIPKIP, Kabillion, Kablammo, Kafziel, Kaihsu, Kaiwhakahaere, Kala25, Kanabekobaton, Karimarie, Karl-Henner, Karmafist, Kathryn NicDhàna, Katieh5584, Kaveh, Kaybull, Kbdank71, Kbh3rd, Keilana, Keith Edkins, Keithdunwoody, KelvSYC, Kenaldinho10, Kernunnos, Kevin B12, Kevin9164, KevinTR, Kevinmon, Kevinzhermin, Keyrocks, Kfcdesuland, Kfitzgib, Khoikhoi, Kidsrus, Kikinou, Kimchi.sg, King Toadsworth, King konger, King of Hearts (old account 2), Kingboyk, Kingjeff, Kingmahad121, Kinigi, Kixix, Kizara, Kizor, Kkm010, Kkornack, Klepto909, Kmccoy, Kneiphof, KnowledgeOfSelf, Kobbra, Kobebryantthejr, Konstable, Kornerkick29, Kosebamse, Kowtoo, Kpalion, Kri, Krich, Kruckenberg.1, KsprayDad, Kubigula, Kukini, Kungfuadam, Kungming2, Kuru, Kurykh, Kwekubo, Kwesiidun91, Kwiki, Kylekyle1, Kylev1, KyraVixen, L Kensington, LSehy, Lacrimosus, Laenfant, Lance181, Landonbrowne, Lardman, Latics, Laughing Man, Laughingirl1515, Lazerburn33, Lazycouch2, Leafyplant, Leandrod, Lectonar, Lee Gregz, LegolasGreenleaf, Leminy, Lemonus, Leon7, Lerdthenerd, Leszek Jańczuk, Liebitz, Liftarn, Lightdarkness, Lightmouse, Lights, Lilcou123, Lilyana, Lindberg, Linkspamremover, Linuxerist, Lion King, Littlechip90, Lkkn, Llcoolspence, Loboxx36, Lofike, Logan, Loganbking, LonelyMarble, Lord Hawk, Lord Marco, Lord Voldemort, LordRM, LordSimonofShropshire, Lordkazan, Loren36, Lost4eva, Lotje, Lowercase, Lowzeewee, Lpw, LuFisto, Lucas B. de Souza, Lucy-marie, Ludraman, LukeSurl, Lumpy joe108, Luna Santin, Lupin, M baptiste, MECU, MER-C, MJCdetroit, MK8, MNAdam, MZMcBride, Mac, Mac Gille Domhnaich, Mac12, MacGyverMagic, Mad sam, MadGeographer, Maday94, Madchester, Maddox, Madw, Maestro25, Magic in the night, Magister Mathematicae, Magooch2.0, Mah favourite, Mahahahaneapneap, Mahanga, Maihu don, Maitch, Majorly, Malhonen, Malo, Malpass93, Mandel, Maniacgeorge, Maniwar, Manuel11223, Manutd 12345, MapsMan, Marc Venot, Marcika, Marco.spiro, Mareino, Marek69, Marianocecowski, Mark, Mark272, MarkA12, Marrilpet, Marshman, MartinHarper, MartinVillafuerte85, Martingio15, Martyn Smith, Marudubshinki, Master Of Ninja, Masterpjz9, Materialscientist, Mateus RM, Matggggg, Mathnarg, Matiasmoreno, Matt Yeager, Matt r kelly, Matthew R Dunn, Matthewfreile, Mattisse, Matty2002, Mauro100, Mav, Maveric56, Maw, Maxamegalon2000, Maxim, Maximaximax, Maxtin, Mayumashu, Mbecker, McDogm, McGeddon, McKennaMan929, Mcdeans, Mdr2375, Meb83, Meelar, Megan1967, Mendor, Meow1234, Merrettrg01, Methnor, Metropolitan90, Mets501, MetsFan76, Mgaigg, Mglovesfun, Michael (vandal), Michael Hardy, Michael Hawkins, Michael Hinrichs, Michael Johnson, Michael Zimmermann, Michael riber jorgensen, Michael-Zero, MickMacNee, MickWest, Midgrid, Miguel.mateo, Mike Rosoft, Mike jones gyeah, Mike1, MikeDog94, Mikethechicken, Mikm, Miles Blues, Minesweeper, Minna Sora no Shita, Mintguy, Mirage5000, MisfitToys, Mistakefinder, Mistamagic28, MisterSheik, Misure poo poo, Misza13, Mkamensek, Mmoneypenny, Mmortal03, Mnpeter, MoRsE, Mochachocca, Moeron, Moeyman, Mohammad adil, Monarchius, Money Makers R Us, Monkbel, Monkeyman, Monkeyman11, Mono, Mopyu90, Moravice, Mormegil, Mothani90, Motley Crue Rocks, Mourn, Moyogo, Mr Chuckles, Mr. Bouncy, Mr. G. Williams, MrFish, MrH, MrJanitor1, MrTranscript, Mreult, Mroach12, Mrxcol, Ms2ger, Mshizzi, Mswake, Muchness, Muhammad Mahdi Karim, Mushroom, Mxalienraptor, Mxcatania, Myanw, Mybighead, Mysdaao, Mysekurity, N458jhk1, Nabla, Naddy, Nakon, Nanonic, Narco, Narsamson, Narssarssuaq, Nascar1996, Nath1012, Natjenko, NawlinWiki, Nawsum526, Ncmvocalist, Nebabc11, Nebu Pookins, Nedarb0, NeferSnoopy, Neil Leslie, Neiman002, New Vandal account!, New World Man, NewEnglandYankee, Newberygreen1, Newone, Nhilary, Nichalp, Nick, Nick Boulevard, Nick of nickness in Colorado, NickCT, Nickshanks, Night Gyr, Nihiltres, Nitaro02, Niteowlneils, Nmitbutcher, No Guru, NocturneNoir, Nodoremi, Noeticsage, Noirceuil, Nojika, Nolitafairytale, Nonagonal Spider, Nonforma, Norbutt2001, Northumbrian, Nosskyline, Nrbelex, Nscheffey, Nsh, Nuggets, Num1dgen, Number 57, Nzd, Nzfooty, OLDMAN, Obamafan70, Obli, Ocaasi, Ocrasaroon, Off!, Off2riorob, Ohconfucius, Ohheyyyy, Ohnoitsjamie, Oldelpaso, Oldhamlet, Olivier, Ollytedwards, Olorin28, Omegacarno07, Omglol, Omicronpersei8, On Thermonuclear War, One, OneManDown, OoberMick, Oogadabaga, Orangemaster, Ordinaria, Ortolan88, Ospalh, Othersider, Ouishoebean, Overlord pat, Ownage24, Oxfordwang, Ozbandit, PHDrillSergeant, PIO, Paavo273, Pacheto, Page Up, Pak21, Paleorthid, Palma234@sympatico.ca, Palmiped, Panarjedde, Papitasfritas, Parable1991, Paris 16, Park3r, Parksand, Parkwells, Pat Gibson, Patayal, Patchamo123, Patrick, Patrick-br, Paul August, Pb30, Pdcook, Peanut4, PeeJay2K3, Persian Poet Gal, Perstar, Peruvianllama, Pespilaludo, Peter Eedy, Peter Isotalo, Peter S., PeterGrecian, Peterkoning, Peterstannard, Pewwer42, Phantomsnake, Phil Boswell, Philip Baird Shearer, PhilipO, PhilipR, Phillip J, Phobal, Phoenix2, PiRSquared17, Picapica, Pieguy48, PierceCheng, Pilotboy5, Pilotguy, Pinircc65, Piontek, Pippu d'Angelo, Pit, Pizzaguy4378, Pladask, Plasticup, Plastikspork, Playak, Plm209, Poborak, PoeticVerse, Poindexter Propellerhead, Pokadoteliza, Pol098, Polarman, Poobslag, Poochy, Porqin, Portiere 101, Portillo, Poslanik, PotentialDanger, Poulsen, Powelldinho, Pparazorback, Ppntori, Pratj, PrestonH, Pretty Green, Prewitt81, PrimeCupEevee, Prodego, Pruneau, PseudoSudo, Psveindhoven, Psycho Kirby, Pusht, Python eggs, QAQUAU, Qrc2006, Quadell, Quae legit, QueenCake, Quilker, Quintillion, Quinxorin, QuizQuick, QwerpQwertus, R Lowry, RB972, RHaworth, RJSampson, RPlunk2853, RabidWalrus, Radio Guy, Raelx, Raf45Martinez, Rafaelamonteiro80, Rahuloof, Raichu, Rainjar, Rak-Tai, Rampinofrancesco, Ranjithsutari, Rapido, Raul654, Raven4x4x, Ravik, Ray Radlein, Ray7jd, Rd232, Rdikeman, Rdsmith4, Readeraml86, ReadingOldBoy, Recury, Red Director, Redcongocross, Reddi, Redrocketboy, Redshoeszeff, Redsox7897, Redvers, Redwolf24, Reeinstein, Refsworldlee, Regentagger, Reggy123, Reginmund, Regular Daddy, Renegadeviking, Rettetast, Rex the first, RexNL, Reyfan1710, Reywas92, Ricardo Moog, Ricardo monteiro, Rich Farmbrough, Richard Allen, Richard Harvey, Richard Rundle, Rick.G, RickK, Ricky81682, Rickyrab, Rjwilmsi, Rkstafford, Rlynagh-shannon, Rmachenw, Rmhermen, Robdurbar, RobertG, Roberta F., Robertgreer, Robversion1, Robwingfield, Rocket71048576, Rocketboy50, Roger491127, Rogerzilla, Romanm, Ronald Mexico, Rory096, Rosiethegreat, RoyBoy, Royalguard11, Royboycrashfan, Rreagan007, Rulesfan, Rx23xaexstlx, Ryan's mom, Ryan25, Rynne, Ryoissoawesome, Ryulong, S0ccerstud301, S0me l0ser, SAFCjl, SFC9394, SJK, SJP, SM, SMcCandlish, SNIyer12, SQGibbon, Sadettin, Sagz639, Salmon, Salt Yeung, Salvio giuliano, SalvoCalcio, Sam Hocevar, Sam Vimes, Samelchami, Sammalin, Sammyjames, Samsara, Samuel Blanning, SandyGeorgia, Sango123, Sangyerians, SaraBaugh, Sasajid, Sasquatch, Sausage948, Sbluen, Sbrools, Sceptre, SchfiftyThree, Schluum, Schrei, Scienter, Sciurinæ, Scottster03, Sdornan, Sean Whitton, Seanchelsea5, Searlewd, Sebasbronzini, Sebastiankessel, Seglea, Seidenstud, Serein (renamed because of SUL), Serminigo, Serpent-A, Serte, Sethlopez, Sfahey, Sgsilver, Shaddens ate denzel for brekky, Shadowjams, Shaidar cuebiyar, Shanes, Shannon bohle, Shirimasen, Shizhao, Shoemakerenator, Shojeeb, Shortenfs, Shotwell, Showersap12, Shrine of Fire, Shushruth, SidP, Sideshow Bob, Siebren, Sietse Snel, Silence, Silivrenion, SilkTork, Sillyfolkboy, Simba1409, SimonMayer, SimonP, SineWave, Sinteractive, Siqbal, Sir Cumsize, Sir Nicholas de Mimsy-Porpington, SirChan, SirGrant, SirJibby, Sirkad, Sixest, Sjakkalle, Sjorford, Sjö, Skh, Skinnyweed, SkullWeasel, Sky83, Skyler13, Slakr, Slow Graffiti, Sluj, Slumgum, Smart7167126716, Smaug123, Smithsterj, SmthManly, Snarfsnarf, Snipeboy09, Snori, SoCalSuperEagle, SoLando, Soaringbear, Sobolewski, Soccer buff, Soccer plyr66, Soccer219, SoccerNews, Soccerdude, Socceredito, Soccerkick8, Soccerking11, Soccerman10103, Soccershoes1, Socialsoccer.com, Sofie310, SofiePedersen, Sonjaaa, Soulresin, SouthernNights, SpNeo, Spaniard78, SparqMan, Speed Air Man, Speedy McG, Spencer, SpencerTC, Speshel k, Spewmaster, SpiceMan, Spike Wilbury, Spitzerman, Spliffy, Splintax, Springeragh, SpuriousQ, Spykesinmahshoe, Sq178pv, Squall88, Srose, Ssilvers, St.daniel, Stalfur, Stanrosander, Stanza13, Starcircloco, Starx, Statistic94, Steel, Steers82, Steinan3, Stephen G. Brown, Stephen Parnell, Stephenb, Stepshep, SteveSims, Stevekeiretsu, Stevenhonig, Stevertigo, Stevey7788, Stevo1000, Stockdiver, Stop Climate Change, Struway, Struway2, Stu.W UK, Studerby, Stui, Stylus Happenstance, Stymphal, Suhalbansal, Suisui, Summerford40, SupaStarGirl, Supadawg, Supplementary, Svick, Swainstonation, Swanyk, Symane, Symon, SynergyBlades, Szater, TBrandley, TFCforever, THUGCHILDz, TLE, TNB774, TRBlom, TSO1D, Taamu, Tad Lincoln, Taemyr, Tainter, TakuyaMurata, Tameamseo, Tanaats, Tancred, Tangerines, Tannin, Tanv91, Tara969, Taroaldo, Tasc, Tata11, Tavix, Tawker, Taylorr, Tbhotch, TeaDrinker, Teapeat, Tedcurly, Tellyaddict, TexMurphy, Thaurisil, The 984, The Anome, The Big C, The Font, The Gnome, The Halo, The Haunted Angel, The Rambling Man, The sock that should not be, The way, the truth, and the light, The wub, TheAmericanizator, TheGrappler, TheProject, Theda, Thefullback, Themodelcitizen, Thenotoriousadin, Therandomthing, Therightclique, Thesoccerjunky4, Thing that goes on feet, Thistle71190, Thryduulf, Thumperward, ThunderPeel2001, Thunderboltz, Tiago Heitor, Tibre, TiffaF, TigerShark, Tim R, Timc, Timmc9106, Timrollpickering, Timwi, Tiresais, Titansfan7532, Titoxd, Toby Woodwark, Tocino, Tom harrison, Tom-3124, Tom-3124(back again), TomCat4680, TomPhil, TomasBat, Tony1,

TonySt, Tonywalton, Toondude, TootinDaylight, Totsie07, Tpbradbury, Travelbird, Tree Biting Conspiracy, Treed, Trevor MacInnis, Trieste, Trixxy, Trontonian, Trosk, Trouserswonky, Troutrooper, Trovatore, Trusilver, TurboGUY, Tvaughn05, TwoOneTwo, Tyciol, Tyomitch, Uannis, UberCryxic, Ughs, Ultimus, Ultratomio, Un chien andalou, Une fette, Unexplained, Unibond, UnicornTapestry, Unreal7, Uranium grenade, UrbanNerd, Uris, Urzadek, Usergreatpower, Utcursch, Utopial, V111P, VMohanty, Vanjagenije, Vanky, Varitek, Vary, Veed, Vega84, VegaDark, Veinor, Venu62, Vf7993, Viajero, Vidshow, Vikreykja, Vilcxjo, VinceBowdren, Violetriga, Viridae, Viriditas, Vivio Testarossa, Vrenator, VsD, Vzbs34, W2ch00, WATP, WAvegetarian, Wackywace, Waggers, WaitingForConnection, Wallerstein, Warfvinge, Warp457, Waseh123321, Wathiik, Wavelength, Wayward, Weregerbil, West coast, Westendgirl, Whall2004, Whatupnathan, Where, Whisky drinker, WhisperToMe, Whouk, Wi-king, Wiki Raja, Wiki alf, Wikibofh, Wikieditor06, Wikien2009, Wikitanvir, Wildnox, Wildplayer61, Will Beback Auto, Willbellum, Willis, Willy on WheeLs, Willy turner, Wingsandsword, Winhunter, Winston365, Wipe, Wknight94, Wmahan, WojPob, Wolfmankurd, Wonderwizard, WoodenTaco, Woody, Woohookitty, Wrp103, Wstitans11, Wtmitchell, Wtwilson3, Wutizevrybudylookingat?, Ww2censor, X!, X201, Xanthoxyl, Xaosflux, Xezbeth, Xhack, Xiaodai, Xiner, Xinit, Xornok, Xtra, Xzyilum, Y, Yarnalgo, Yaz0r, Yclept:Berr, Yeahsoo, Yekrats, YellowMonkey, Yoenit, Yonatan, Youndbuckerz, Your soccer boy, Yousou, Ysangkok, Yuckfoo, YuffieTheBear, Zachary Allen Berg, Zahid Abdassabur, Zanimum, Zaphod Beeblebrox, Zaslav, Zephyr21, Zhelja, ZimZalaBim, Zizouthecat, Zocky, Zolstijers, Zreeon, Zundark, Zytroft, Zzmonty, Zzuuzz, Zzyzx11, [212], Ævar Arnfjörð Bjarmason, Александър, Саша Стефановић, ||ممیورری, فرش ا|دلين, 3260 anonymous edits

Image Sources, Licenses and Contributors

File:Kvkfanpages67.jpg *Source*: http://en.wikipedia.org/w/index.php?title=File:Kvkfanpages67.jpg *License*: unknown *Contributors*: User:Giovanny Saelens

Image:Charleroi.gif *Source*: http://en.wikipedia.org/w/index.php?title=File:Charleroi.gif *License*: unknown *Contributors*: Julien Tuerlinckx, LimoWreck, Nanonic, Skier Dude

File:Flag of Iran.svg *Source*: http://en.wikipedia.org/w/index.php?title=File:Flag_of_Iran.svg *License*: unknown *Contributors*: Various

File:Flag of Hungary.svg *Source*: http://en.wikipedia.org/w/index.php?title=File:Flag_of_Hungary.svg *License*: unknown *Contributors*: User:SKopp

File:Flag of Croatia.svg *Source*: http://en.wikipedia.org/w/index.php?title=File:Flag_of_Croatia.svg *License*: unknown *Contributors*: User:MaGa

File:Flag of France.svg *Source*: http://en.wikipedia.org/w/index.php?title=File:Flag_of_France.svg *License*: unknown *Contributors*: Anomie

File:Flag of Romania.svg *Source*: http://en.wikipedia.org/w/index.php?title=File:Flag_of_Romania.svg *License*: unknown *Contributors*: User:AdiJapan

File:Flag of Israel.svg *Source*: http://en.wikipedia.org/w/index.php?title=File:Flag_of_Israel.svg *License*: unknown *Contributors*: AnonMoos, Bastique, Bobika, Brown spite, Captain Zizi, Cerveaugenie, Drork, Etams, Fred J, Fry1989, Geagea, Himasaram, Homo lupus, Humus sapiens, Klemen Kocjancic, Komando12, Kookaburra, Luispihormiguero, Madden, Neq00, NielsF, Nightstallion, Oren neu dag, Patstuart, PeeJay2K3, Pumbaa80, Ramiy, Reisio, Rodejong, SKopp, Sceptic, SomeDudeWithAUserName, Technion, Typhix, Valentinian, Yellow up, Zscout370, 31 anonymous edits

File:Flag of Turkey.svg *Source*: http://en.wikipedia.org/w/index.php?title=File:Flag_of_Turkey.svg *License*: unknown *Contributors*: User:Dbenbenn

File:Flag of Slovakia.svg *Source*: http://en.wikipedia.org/w/index.php?title=File:Flag_of_Slovakia.svg *License*: unknown *Contributors*: User:SKopp

File:Flag of England.svg *Source*: http://en.wikipedia.org/w/index.php?title=File:Flag_of_England.svg *License*: unknown *Contributors*: Anomie

File:Flag of Denmark.svg *Source*: http://en.wikipedia.org/w/index.php?title=File:Flag_of_Denmark.svg *License*: unknown *Contributors*: User:Madden

File:Flag of Wales 2.svg *Source*: http://en.wikipedia.org/w/index.php?title=File:Flag_of_Wales_2.svg *License*: unknown *Contributors*: AlexD, Cecil, Dbenbenn, Duduziq, F. F. Fjodor, FruitMonkey, Fry1989, Homo lupus, Iago4096, IvanOS, Pumbaa80, Red devil 666, Srtxg, Tha real, Torstein, Vernanimalcula, Vzb83, Wouterhagens, 7 anonymous edits

File:Flag of Poland.svg *Source*: http://en.wikipedia.org/w/index.php?title=File:Flag_of_Poland.svg *License*: unknown *Contributors*: Anomie, Mifter

File:Flag of Austria.svg *Source*: http://en.wikipedia.org/w/index.php?title=File:Flag_of_Austria.svg *License*: unknown *Contributors*: User:SKopp

File:Flag of Finland.svg *Source*: http://en.wikipedia.org/w/index.php?title=File:Flag_of_Finland.svg *License*: unknown *Contributors*: User:SKopp

File:Flag of Slovenia.svg *Source*: http://en.wikipedia.org/w/index.php?title=File:Flag_of_Slovenia.svg *License*: unknown *Contributors*: User:SKopp, User:Vzb83, User:Zscout370

File:Flag of Spain.svg *Source*: http://en.wikipedia.org/w/index.php?title=File:Flag_of_Spain.svg *License*: unknown *Contributors*: Anomie

File:Flag of Belgium (civil).svg *Source*: http://en.wikipedia.org/w/index.php?title=File:Flag_of_Belgium_(civil).svg *License*: unknown *Contributors*: Bean49, David Descamps, Dbenbenn, Denelson83, Evanc0912, Fry1989, Gabriel trzy, Howcome, IvanOS, Ms2ger, Nightstallion, Oreo Priest, Rocket000, Rodejong, Sir Iain, ThomasPusch, Warddr, Zscout370, 4 anonymous edits

File:Flag of the Democratic Republic of the Congo.svg *Source*: http://en.wikipedia.org/w/index.php?title=File:Flag_of_the_Democratic_Republic_of_the_Congo.svg *License*: unknown *Contributors*: Anime Addict AA, AnonMoos, Antemister, ChongDae, Courcelles, Denelson83, Ed veg, FischersFritz, Fry1989, Future Perfect at Sunrise, Homo lupus, Klemen Kocjancic, LA2, Mattes, Moyogo, Nagy, Neq00, Nightstallion, ThomasPusch, Urhixidur, Vzb83, Wester, Zscout370, 11 anonymous edits

File:Flag of Ghana.svg *Source*: http://en.wikipedia.org/w/index.php?title=File:Flag_of_Ghana.svg *License*: unknown *Contributors*: Benchill, Fry1989, Henswick, Homo lupus, Indolences, Jarekt, Klemen Kocjancic, Magasjukur2, Neq00, OAlexander, SKopp, ThomasPusch, Threecharlie, Torstein, Zscout370, 5 anonymous edits

File:Flag of Senegal.svg *Source*: http://en.wikipedia.org/w/index.php?title=File:Flag_of_Senegal.svg *License*: unknown *Contributors*: user:Nightstallion

File:Flag of Montenegro.svg *Source*: http://en.wikipedia.org/w/index.php?title=File:Flag_of_Montenegro.svg *License*: unknown *Contributors*: User:B1mbo, User:Froztbyte

File:Flag of Cameroon.svg *Source*: http://en.wikipedia.org/w/index.php?title=File:Flag_of_Cameroon.svg *License*: unknown *Contributors*: (of code)

File:Flag of Brazil.svg *Source*: http://en.wikipedia.org/w/index.php?title=File:Flag_of_Brazil.svg *License*: unknown *Contributors*: Anomie

File:Flag of Germany.svg *Source*: http://en.wikipedia.org/w/index.php?title=File:Flag_of_Germany.svg *License*: unknown *Contributors*: Anomie

File:Flag of Switzerland.svg *Source*: http://en.wikipedia.org/w/index.php?title=File:Flag_of_Switzerland.svg *License*: unknown *Contributors*: User:-xfi-, User:Marc Mongenet, User:Zscout370

File:Flag of Armenia.svg *Source*: http://en.wikipedia.org/w/index.php?title=File:Flag_of_Armenia.svg *License*: unknown *Contributors*: User:SKopp

File:Flag of Algeria.svg *Source*: http://en.wikipedia.org/w/index.php?title=File:Flag_of_Algeria.svg *License*: unknown *Contributors*: User:SKopp

File:Flag of Argentina.svg *Source*: http://en.wikipedia.org/w/index.php?title=File:Flag_of_Argentina.svg *License*: unknown *Contributors*: User:Dbenbenn

File:Flag of Australia.svg *Source*: http://en.wikipedia.org/w/index.php?title=File:Flag_of_Australia.svg *License*: unknown *Contributors*: Anomie, Mifter

File:Flag of Bosnia and Herzegovina.svg *Source*: http://en.wikipedia.org/w/index.php?title=File:Flag_of_Bosnia_and_Herzegovina.svg *License*: unknown *Contributors*: User:Kseferovic

File:Flag of Burkina Faso.svg *Source*: http://en.wikipedia.org/w/index.php?title=File:Flag_of_Burkina_Faso.svg *License*: unknown *Contributors*: User:Gabbe, User:SKopp

File:Flag of Chile.svg *Source*: http://en.wikipedia.org/w/index.php?title=File:Flag_of_Chile.svg *License*: unknown *Contributors*: Alkari, B1mbo, David Newton, Dbenbenn, Denelson83, ElmA, Er Komandante, Fibonacci, Fry1989, Fsopolonezcaro, Herbythyme, Huhsunqu, Kallerna, Kanonkas, Klemen Kocjancic, Kyro, Mattes, Mozzan, Nagy, Nightstallion, Piastu, Pixeltoo, Pumbaa80, SKopp, Sarang, Srtxg, Sterling.M.Archer, Str4nd, Ultratomio, Vzb83, Xarucoponce, Yakoo, Yonatanh, Zscout370, 42 anonymous edits

File:Flag of Guinea.svg *Source*: http://en.wikipedia.org/w/index.php?title=File:Flag_of_Guinea.svg *License*: unknown *Contributors*: User:SKopp

File:Flag of Ireland.svg *Source*: http://en.wikipedia.org/w/index.php?title=File:Flag_of_Ireland.svg *License*: unknown *Contributors*: User:SKopp

File:Flag of Luxembourg.svg *Source*: http://en.wikipedia.org/w/index.php?title=File:Flag_of_Luxembourg.svg *License*: unknown *Contributors*: User:SKopp

File:Flag of Macedonia.svg *Source*: http://en.wikipedia.org/w/index.php?title=File:Flag_of_Macedonia.svg *License*: unknown *Contributors*: User:Gabbe, User:SKopp

File:Flag of Morocco.svg *Source*: http://en.wikipedia.org/w/index.php?title=File:Flag_of_Morocco.svg *License*: unknown *Contributors*: User:Denelson83, User:Zscout370

File:Flag of the Netherlands.svg *Source*: http://en.wikipedia.org/w/index.php?title=File:Flag_of_the_Netherlands.svg *License*: unknown *Contributors*: User:Zscout370

File:Flag of Nigeria.svg *Source*: http://en.wikipedia.org/w/index.php?title=File:Flag_of_Nigeria.svg *License*: unknown *Contributors*: User:Jhs

File:Flag of Serbia.svg *Source*: http://en.wikipedia.org/w/index.php?title=File:Flag_of_Serbia.svg *License*: unknown *Contributors*: sodipodi.com

File:Flag of Sierra Leone.svg *Source*: http://en.wikipedia.org/w/index.php?title=File:Flag_of_Sierra_Leone.svg *License*: unknown *Contributors*: Zscout370

File:Flag of Sweden.svg *Source*: http://en.wikipedia.org/w/index.php?title=File:Flag_of_Sweden.svg *License*: unknown *Contributors*: Anomie

File:Flag of Togo.svg *Source*: http://en.wikipedia.org/w/index.php?title=File:Flag_of_Togo.svg *License*: unknown *Contributors*: Aaker, Ahsoous, Alkari, EugeneZelenko, Fry1989, Homo lupus, Klemen Kocjancic, Mattes, Mxn, Neq00, Nightstallion, Reisio, ThomasPusch, Vzb83

File:Flag of Ukraine.svg *Source*: http://en.wikipedia.org/w/index.php?title=File:Flag_of_Ukraine.svg *License*: unknown *Contributors*: User:Jon Harald Søby, User:Zscout370

File:Flag of Oman.svg *Source*: http://en.wikipedia.org/w/index.php?title=File:Flag_of_Oman.svg *License*: unknown *Contributors*: *drew, Duduziq, Fry1989, Happenstance, Homo lupus, Ittihadawi, Jetijones, Klemen Kocjancic, Liftarn, Mattes, Neq00, Nightstallion, NikNaks, OAlexander, Orange Tuesday, Pumbaa80, Rfc1394, ThomasPusch, Zscout370

File:Flag of Scotland.svg *Source*: http://en.wikipedia.org/w/index.php?title=File:Flag_of_Scotland.svg *License*: unknown *Contributors*: User:Kbolino

File:Flag of Russia.svg *Source*: http://en.wikipedia.org/w/index.php?title=File:Flag_of_Russia.svg *License*: unknown *Contributors*: Anomie

File:Flag of Greece.svg *Source*: http://en.wikipedia.org/w/index.php?title=File:Flag_of_Greece.svg *License*: unknown *Contributors*: (of code) (talk)

File:Flag of Portugal.svg *Source*: http://en.wikipedia.org/w/index.php?title=File:Flag_of_Portugal.svg *License*: unknown *Contributors*: User:Nightstallion

File:Flag of Iceland.svg *Source*: http://en.wikipedia.org/w/index.php?title=File:Flag_of_Iceland.svg *License*: unknown *Contributors*: User:Zscout370, User:Ævar Arnfjörð Bjarmason

File:Flag of the Faroe Islands.svg *Source*: http://en.wikipedia.org/w/index.php?title=File:Flag_of_the_Faroe_Islands.svg *License*: unknown *Contributors*: User:IceKarma

File:Flag of Georgia.svg *Source*: http://en.wikipedia.org/w/index.php?title=File:Flag_of_Georgia.svg *License*: unknown *Contributors*: User:SKopp

File:Flag of Albania.svg *Source*: http://en.wikipedia.org/w/index.php?title=File:Flag_of_Albania.svg *License*: unknown *Contributors*: User:Dbenbenn

File:Flag of Estonia.svg *Source*: http://en.wikipedia.org/w/index.php?title=File:Flag_of_Estonia.svg *License*: unknown *Contributors*: User:PeepP, User:SKopp

File:Flag of Cote d'Ivoire.svg *Source*: http://en.wikipedia.org/w/index.php?title=File:Flag_of_Cote_d'Ivoire.svg *License*: unknown *Contributors*: User:Jon Harald Søby

File:Flag of Tunisia.svg *Source*: http://en.wikipedia.org/w/index.php?title=File:Flag_of_Tunisia.svg *License*: unknown *Contributors*: Alkari, AnonMoos, Avala, Bender235, Duduziq, Elina2308, Emmanuel.boutet, Flad, Fry1989, Gabbe, Juiced lemon, Klemen Kocjancic, Mattes, Meno25, Moumou82, Myself488, Neq00, Nightstallion, Reisio, Str4nd, TFCforever, Ö, Фёдор Гусляров, 9 anonymous edits

File:Flag of Norway.svg *Source*: http://en.wikipedia.org/w/index.php?title=File:Flag_of_Norway.svg *License*: unknown *Contributors*: User:Dbenbenn

File:Flag of South Africa.svg *Source*: http://en.wikipedia.org/w/index.php?title=File:Flag_of_South_Africa.svg *License*: unknown *Contributors*: Adriaan, Anime Addict AA, AnonMoos, BRUTE, Daemonic Kangaroo, Dnik, Duduziq, Dzordzm, Fry1989, Homo lupus, Jappalang, Juliancolton, Kam Solusar, Klemen Kocjancic, Klymene, Lexxyy, Mahahahaneapneap, Manuelt15, Moviedefender, NeverDoING, Ninane, Poznaniak, Przemub, SKopp, ThePCKid, ThomasPusch, Tvdm, Ultratomio, Vzb83, Zscout370, 35 anonymous edits

File:Flag of Niger.svg *Source*: http://en.wikipedia.org/w/index.php?title=File:Flag_of_Niger.svg *License*: unknown *Contributors*: User:verdy_p

File:Flag of the Czech Republic.svg *Source*: http://en.wikipedia.org/w/index.php?title=File:Flag_of_the_Czech_Republic.svg *License*: unknown *Contributors*: special commission (of code): SVG version by cs:-xfi-. Colors according to Appendix No. 3 of czech legal Act 3/1993. cs:Zirland.

File:Flag of SFR Yugoslavia.svg *Source*: http://en.wikipedia.org/w/index.php?title=File:Flag_of_SFR_Yugoslavia.svg *License*: unknown *Contributors*: Flag designed by Đorđe Andrejević-KunSVG: Zscout370 at en.wikipedia

Image:Zultewaregemlogo.png *Source*: http://en.wikipedia.org/w/index.php?title=File:Zultewaregemlogo.png *License*: unknown *Contributors*: Ief, Nanonic, Papado2, Skier Dude

Image:Newcastle United v Zulte Waregem, 2007 (4).JPG *Source*: http://en.wikipedia.org/w/index.php?title=File:Newcastle_United_v_Zulte_Waregem,_2007_(4).JPG *License*: unknown *Contributors*: User:Responsible?

File:Flag of the Central African Republic.svg *Source*: http://en.wikipedia.org/w/index.php?title=File:Flag_of_the_Central_African_Republic.svg *License*: unknown *Contributors*: User:Nightstallion

File:Flag of Belarus.svg *Source*: http://en.wikipedia.org/w/index.php?title=File:Flag_of_Belarus.svg *License*: unknown *Contributors*: User:Zscout370

File:Flag of Peru.svg *Source*: http://en.wikipedia.org/w/index.php?title=File:Flag_of_Peru.svg *License*: unknown *Contributors*: User:Dbenbenn

File:Arsenal 1886-2011 Logo.png *Source*: http://en.wikipedia.org/w/index.php?title=File:Arsenal_1886-2011_Logo.png *License*: unknown *Contributors*: Fadzrulhisyam, Malpass93, Salvio giuliano, Sfan00 IMG, Zzizoo, 1 anonymous edits

Image:Woolwich Arsenal v. Newcastle United, April 1906.jpg *Source*: http://en.wikipedia.org/w/index.php?title=File:Woolwich_Arsenal_v._Newcastle_United,_April_1906.jpg *License*: unknown *Contributors*: J 1982, Qwghlm, Responsible?

File:Arsenal open top bus parade 2004.jpg *Source*: http://en.wikipedia.org/w/index.php?title=File:Arsenal_open_top_bus_parade_2004.jpg *License*: unknown *Contributors*: Qwghlm, Thuresson, 2 anonymous edits

File:Arsenal FC league finishing positions since 1947.png *Source*: http://en.wikipedia.org/w/index.php?title=File:Arsenal_FC_league_finishing_positions_since_1947.png *License*: unknown *Contributors*: User:Crookesmoor

File:Arsenal crest 1888.png *Source*: http://en.wikipedia.org/w/index.php?title=File:Arsenal_crest_1888.png *License*: unknown *Contributors*: Ed g2s, Notwist, Xenophon, 2 anonymous edits

File:Arsenal Crest 1930.svg *Source*: http://en.wikipedia.org/w/index.php?title=File:Arsenal_Crest_1930.svg *License*: unknown *Contributors*: own work

File:Arsenal Crest 1952.svg *Source*: http://en.wikipedia.org/w/index.php?title=File:Arsenal_Crest_1952.svg *License*: unknown *Contributors*: own work

File:Arsenal fc old crest small.png *Source*: http://en.wikipedia.org/w/index.php?title=File:Arsenal_fc_old_crest_small.png *License*: unknown *Contributors*: Original uploader was Ed g2s at en.wikipedia Later versions were uploaded by Mahahahaneapneap at en.wikipedia.

File:Arsenal FC.svg *Source*: http://en.wikipedia.org/w/index.php?title=File:Arsenal_FC.svg *License*: unknown *Contributors*: Avicennasis, BigDom, Drilnoth, Ed g2s, Flrn, Gump Stump, Lord Pistachio, Malpass93, Minimarcus, Ngckmax, Pais, Polly, Qwghlm, Salvio giuliano, Skier Dude, 8 anonymous edits

Image:Arsenal Stadium interior North Bank.jpg *Source*: http://en.wikipedia.org/w/index.php?title=File:Arsenal_Stadium_interior_North_Bank.jpg *License*: unknown *Contributors*: Bhoeble, Boscorelli, Ixfd64, Jonathan Cardy, Qwghlm, 3 anonymous edits

Image:Emirates Stadium Arsenal.jpg *Source*: http://en.wikipedia.org/w/index.php?title=File:Emirates_Stadium_Arsenal.jpg *License*: unknown *Contributors*: Original uploader was Arsenal mjh at en.wikipedia

File:Laurent Koscielny clashes with Heurelho Gomes.jpg *Source*: http://en.wikipedia.org/w/index.php?title=File:Laurent_Koscielny_clashes_with_Heurelho_Gomes.jpg *License*: unknown *Contributors*: Ronnie Macdonald

File:Flag of South Korea.svg *Source*: http://en.wikipedia.org/w/index.php?title=File:Flag_of_South_Korea.svg *License*: unknown *Contributors*: Various

File:Flag of Jamaica.svg *Source*: http://en.wikipedia.org/w/index.php?title=File:Flag_of_Jamaica.svg *License*: unknown *Contributors*: Anime Addict AA, Boricuaeddie, Davepape, Duduziq, Fred J, Fry1989, Herbythyme, KBarnett, Kilom691, Klemen Kocjancic, Kounoupidi, Körnerbrötchen, Ludger1961, Mattes, Nishkid64, Odder, Reisio, SKopp, Sarang, The Evil IP address, Wknight94, 27 anonymous edits

File:Flag of Italy.svg *Source*: http://en.wikipedia.org/w/index.php?title=File:Flag_of_Italy.svg *License*: unknown *Contributors*: Anomie

File:Flag of Japan.svg *Source*: http://en.wikipedia.org/w/index.php?title=File:Flag_of_Japan.svg *License*: unknown *Contributors*: Anomie

File:Flag of Costa Rica.svg *Source*: http://en.wikipedia.org/w/index.php?title=File:Flag_of_Costa_Rica.svg *License*: unknown *Contributors*: User:Gabbe, User:SKopp

File:Flag of Mexico.svg *Source*: http://en.wikipedia.org/w/index.php?title=File:Flag_of_Mexico.svg *License*: unknown *Contributors*: User:AlexCovarrubias

Image:Arsene Wenger2009.JPG *Source*: http://en.wikipedia.org/w/index.php?title=File:Arsene_Wenger2009.JPG *License*: unknown *Contributors*: Olaf Nordwich Original uploader was ChrisTheDude at en.wikipedia

File:Club Brugge KV.png *Source*: http://en.wikipedia.org/w/index.php?title=File:Club_Brugge_KV.png *License*: unknown *Contributors*: BigrTex, Foxhill, Melesse, Nanonic, Ultras GE

File:Janbreydelstadion.jpg *Source*: http://en.wikipedia.org/w/index.php?title=File:Janbreydelstadion.jpg *License*: unknown *Contributors*: dju22000

File:Clubbruggetifo.jpg *Source*: http://en.wikipedia.org/w/index.php?title=File:Clubbruggetifo.jpg *License*: unknown *Contributors*: dju22000

File:Flag of Colombia.svg *Source*: http://en.wikipedia.org/w/index.php?title=File:Flag_of_Colombia.svg *License*: unknown *Contributors*: User:SKopp

File:Flag of the Republic of the Congo.svg *Source*: http://en.wikipedia.org/w/index.php?title=File:Flag_of_the_Republic_of_the_Congo.svg *License*: unknown *Contributors*: Anime Addict AA, Antemister, Courcelles, Denelson83, Estrilda, FischersFritz, Fry1989, Homo lupus, Klemen Kocjancic, LA2, Madden, Mattes, Moyogo, Neq00, Nightstallion, Persiana, Ratatosk, ThomasPusch, Thuresson, 3 anonymous edits

File:Flag of Czechoslovakia.svg *Source*: http://en.wikipedia.org/w/index.php?title=File:Flag_of_Czechoslovakia.svg *License*: unknown *Contributors*: (of code)

Image:Kortrijk.png *Source*: http://en.wikipedia.org/w/index.php?title=File:Kortrijk.png *License*: unknown *Contributors*: Ief, Salavat, Skier Dude

File:Flag of The Gambia.svg *Source*: http://en.wikipedia.org/w/index.php?title=File:Flag_of_The_Gambia.svg *License*: unknown *Contributors*: Atamari, Avala, Courcelles, Denniss, FischersFritz, Fry1989, Klemen Kocjancic, Mattes, Neq00, Nightstallion, OAlexander, Porao, Rodejong, ThomasPusch, Vzb83, WikipediaMaster, Zscout370, 3 anonymous edits

File:Flag of the Republic of China.svg *Source*: http://en.wikipedia.org/w/index.php?title=File:Flag_of_the_Republic_of_China.svg *License*: unknown *Contributors*: 555, Abner1069, Bestalex, Bigmorr, Denelson83, Ed veg, Gzdavidwong, Herbythyme, Isletakee, Kakoui, Kallerna, Kibinsky, Mattes, Mizunoryu, Neq00, Nickpo, Nightstallion, Odder, Pymouss, R.O.C, Reisio, Reuvenk, Rkt2312, Rocket000, Runningfridgesrule, Samwingkit, Sasha Krotov, Shizhao, Tabasco, Vzb83, Wrightbus, ZooFari, Zscout370, 74 anonymous edits

File:Flag of Bulgaria.svg *Source*: http://en.wikipedia.org/w/index.php?title=File:Flag_of_Bulgaria.svg *License*: unknown *Contributors*: User:SKopp

file:football iu 1996.jpg *Source*: http://en.wikipedia.org/w/index.php?title=File:Football_iu_1996.jpg *License*: unknown *Contributors*: Rick Dikeman

File:U20-WorldCup2007-Okotie-Onka edit2.jpg *Source*: http://en.wikipedia.org/w/index.php?title=File:U20-WorldCup2007-Okotie-Onka_edit2.jpg *License*: unknown *Contributors*: User:Fir0002, User:Nwiebe

File:Soccer goalkeeper.jpg *Source*: http://en.wikipedia.org/w/index.php?title=File:Soccer_goalkeeper.jpg *License*: unknown *Contributors*: U.S. Air Force photo by Master Sgt. Lance Cheung

File:Fußballgeschichte (1872).jpg *Source*: http://en.wikipedia.org/w/index.php?title=File:Fußballgeschichte_(1872).jpg *License*: unknown *Contributors*: Daemonic Kangaroo, FSII, Infrogmation, Juiced lemon, Kinigi, Maksim, Man vyi, Pepito, 3 anonymous edits

File:1stRoyalEngineers.png *Source*: http://en.wikipedia.org/w/index.php?title=File:1stRoyalEngineers.png *License*: unknown *Contributors*: Ffahm, Responsible?

File:Football pitch metric.svg *Source*: http://en.wikipedia.org/w/index.php?title=File:Football_pitch_metric.svg *License*: unknown *Contributors*: User:NielsF

File:Shunsuke1 20080622.jpg *Source*: http://en.wikipedia.org/w/index.php?title=File:Shunsuke1_20080622.jpg *License*: unknown *Contributors*: User:Neier

Image:Yellow card.svg *Source*: http://en.wikipedia.org/w/index.php?title=File:Yellow_card.svg *License*: unknown *Contributors*: User:ed_g2s

Image:Red card.svg *Source*: http://en.wikipedia.org/w/index.php?title=File:Red_card.svg *License*: unknown *Contributors*: User:ed_g2s

File:Ryan Valentine scores.jpg *Source*: http://en.wikipedia.org/w/index.php?title=File:Ryan_Valentine_scores.jpg *License*: unknown *Contributors*: User:Markbarnes

File:Mecz Polska - Armenia 04 ssj 20070328.jpg *Source*: http://en.wikipedia.org/w/index.php?title=File:Mecz_Polska_-_Armenia_04_ssj_20070328.jpg *License*: unknown *Contributors*: User:Staszek_Szybki_Jest

File:Mia1997.JPG *Source*: http://en.wikipedia.org/w/index.php?title=File:Mia1997.JPG *License*: unknown *Contributors*: Johnmaxmena (talk)JOhn Mena. Original uploader was Johnmaxmena at en.wikipedia

MIX
Papier aus verantwortungsvollen Quellen
Paper from responsible sources
FSC® C105338

Printed by Books on Demand GmbH, Norderstedt / Germany